T0327521

International Trade and Agriculture

To our families

Ji, Grace, and John Koo
Julie, John, Sean, and Eryn Kennedy

INTERNATIONAL TRADE AND AGRICULTURE

WON W. KOO AND
P. LYNN KENNEDY

BLACKWELL PUBLISHING
350 Main Street, Malden, MA 02148-5020, USA
108 Cowley Road, Oxford OX4 1JF, UK
550 Swanston Street, Carlton, Victoria 3053, Australia

First published 2005 by Blackwell Publishing Ltd

Library of Congress Cataloging-in-Publication Data
Koo, Won W.
   International trade and agriculture / Won W. Koo and P. Lynn Kennedy.
      p. cm.
   Includes bibliographical references and index.
   ISBN 1-4051-0800-2 (hardback : alk. paper)
   1. International trade.   2. Produce trade.   I. Kennedy, P. Lynn.   II. Title.

   HF1379.K656 2005
   382′.41–dc22

                                                                    2004014015

A catalogue record for this title is available from the British Library.

Set in 10/12.5pt Rotis Serif
by Graphicraft Limited, Hong Kong

For further information on
Blackwell Publishing, visit our website:
www.blackwellpublishing.com

Brief Contents

Contents

Preface and Acknowledgments

The purpose in writing this text is to present theory and principles that are essential for understanding the challenges in international economics; in particular, those issues related to trade of agricultural goods within a rapidly globalizing world. This book is designed to be used over one semester in a course covering both the macro- and micro-components of international economics, primarily at the undergraduate level.

Globalization is a process that has been realized through multilateral and bilateral trade negotiations for the past several decades. Moreover, the rapid development and adoption of technology in recent years has accelerated the process. The ability to instantaneously communicate around the globe has increased the number and breadth of markets for agricultural producers and processors. Advances in transportation and logistics have increased the speed with which agricultural and food products reach their destinations. These developments benefit both producers and consumers through decreased costs and increased efficiency. However, as trade benefits increase, so too do the risks in areas such as food safety and security.

Given the vital role of agricultural and food products in the world economy, it is important to understand international economics as it relates to agriculture. The agricultural industry possesses several unique characteristics that make the study of international agricultural economics critical. For example, the biological characteristics of agricultural and food products require that trade take place in a timely manner. Many other excellent textbooks are available that discuss these issues from a nonagricultural perspective. This book is intended to fill the void by presenting the theory of international economics as it relates to agriculture and then applying the theory to contemporary issues.

To accomplish this objective, the initial part of this text discusses the theory of international economics as it relates to agriculture. Traditional, neoclassical trade theory is presented, in addition to an overview of new trade theory. The second part of the book largely uses a partial equilibrium framework as a tool with which students may analyze and compare the various welfare implications of trade policies and agreements.

Given the increased volume of global agricultural trade, currency exchange rates become more important in determining trade flows. Two chapters are dedicated to an overview of exchange rates, including exchange rate determination and its impact on agricultural trade. The remainder of the book covers topics that are of critical importance in the world economy: trade and development; direct foreign investment; and trade and the environment. While other important issues exist, these three are critical to trade liberalization discussions within the international community. In addition to the theoretical component of the text, we include real-world applications to provide students with a better understanding of actual trade issues and problems that the United States and other countries currently face under a globalizing trade environment.

The process of writing a book of this type cannot be undertaken alone. While we have spent many hours writing and rewriting the chapters contained in this book, there are many individuals who have supported this project. We would like to thank Beth Ambrosio and Jeremy Mattson in the Center for Agricultural Policy and Trade Studies at North Dakota State University (NDSU), and Dae-Seob Lee and Elizabeth Roule in the Department of Agricultural Economics and Agribusiness at Louisiana State University (LSU), for their comments on early drafts of this manuscript. The support of our respective institutions has been tremendous. We would like to thank David Lambert at North Dakota State University, Gail Cramer, L. J. Guedry, and Bill Richardson of the LSU AgCenter, and Ken Koonce of the LSU College of Agriculture, for their encouragement and support throughout this process. While the quality of this text has benefitted from the advice of many reviewers, the authors accept responsibility for any errors or omissions that remain.

Won W. Koo and P. Lynn Kennedy

CHAPTER 1

Introduction

■ 1.0 INTRODUCTION

The world market for agricultural goods allows countries to enhance their welfare through trade. Differences in resource endowments between countries and regions result in each country having unique advantages for producing a particular set of products. Countries can produce the commodities in which they have an advantage and trade for other goods that they wish to consume. For example, one country may be self-sufficient with respect to soybeans, and yet be a large exporter of cotton and an importer of sugar.

Agricultural trade is a unique subset of total merchandise trade. The agricultural industry and agricultural products have many characteristics that distinguish them from other industries. Agricultural goods are often perishable. This makes the timely sale of agricultural products critical. The nature of agricultural production makes it less responsive to short-term price fluctuations. This means that swings in world market prices can have major effects on agricultural income. Agriculture is typically a smaller component of the developed economies than for the less developed world. This implies that uncertainty in the world agricultural market will have a greater impact on less developed countries than on developed countries.

The ability of countries to produce agricultural commodities differs on the basis of resource endowments and technology. The dependence of agriculture on resource endowments results in a degree of specialization in agriculture that is difficult to alter through investment. While capital and skilled labor can be increased through investment and training, the amount of land available for agricultural production is a relatively fixed resource. Given this, countries tend to specialize in the production of agricultural commodities on the basis of their endowment of natural resources. This implies that trading agricultural commodities among nations is critically important to improve utilization of the given natural resources and to optimize production efficiency.

Agricultural trade also plays a unique and varied role in the economic development of nations. Although it may have been an engine of economic growth in the

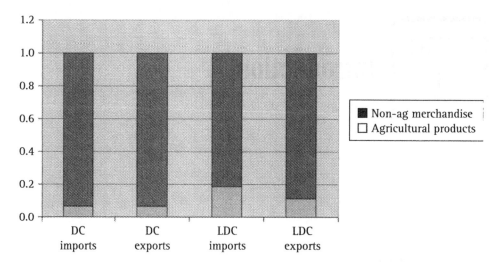

Figure 1.1 The composition of trade in developed (DC) and least developed (LDC) countries as a share of total trade, 2001.
Source: FAOSTAT (2003), United Nations, Rome

early stage of economic development, agriculture is a small portion of the economy in most developed nations. Agriculture's share of total trade is presented for developed and less developed countries in Figure 1.1. In developed countries (DCs), agricultural imports and exports each account for less than 7% of total import and export values, respectively. In the less developed and developing countries (LDCs), agricultural imports are nearly 20% of total imports, while agricultural exports account for 11% of total exports. LDCs are more dependent on agriculture than DCs.

Developed countries, however, tend to provide more protection and support to agriculture than LDCs. In addition to the unique characteristics that set agriculture apart from other industries, agricultural sectors in different countries receive varied support from their respective governments. Agricultural protection tends to insulate domestic markets from the world market. As a result, variability in the world market is magnified. For example, when a large DC imposes an import quota or other import restrictions, supply and demand shocks around the world will affect the world market. This occurs because the import quotas or restrictions act as barriers between markets. In order for the rest of the world to absorb supply and demand shocks, the world price must adjust to a new market-clearing equilibrium. Thus, protectionist policies on the part of DCs destabilize the world market. These shocks impact producers and consumers in countries around the world, but especially in LDCs.

It has been noted that agriculture typically comprises a greater share of production and trade in LDCs than in DCs. Despite this distinction, the world agricultural market is dominated by DCs. Table 1.1 shows data for the largest agricultural importers and exporters in 2001. Most countries export and import agricultural commodities. For some countries, their exports are larger than their imports, resulting in trade surpluses. Other countries import more than they export, resulting in trade deficits.

Table 1.1 The value of total agricultural product trade for the largest importing and exporting countries, 2001 (millions of US dollars)

Country	Agricultural exports	Agricultural imports	Agricultural trade balance
Argentina	10,989	1,210	9,779
Australia	15,779	2,867	12,911
Brazil	16,060	3,209	12,851
Canada	17,270	12,038	5,232
France	31,003	23,224	7,779
Germany	23,582	32,157	−8,574
Italy	15,687	20,915	−5,229
Japan	2,475	34,572	−32,096
Mexico	7,631	10,830	−3,199
Netherlands	27,777	16,978	10,799
Spain	14,504	11,226	3,279
United Kingdom	13,354	26,839	−13,485
United States	56,706	44,940	11,766

Source: FAOSTAT (2003), United Nations, Rome

These data indicate that much of the world's agricultural trade, both imports and exports, occurs among large industrial nations. These countries tend to protect agriculture. For example, in many DCs, agricultural production is subsidized by the government, agricultural exports are encouraged through government supported programs, and agricultural imports are restricted through various mechanisms. This protection serves to insulate the domestic market from world price fluctuations. Many LDCs behave differently from DCs, taxing agricultural exports as a means of increasing government revenues. This contributes to instabilities on the world market.

In an attempt to reduce protection levels and encourage free trade in world markets, the General Agreement on Tariffs and Trade (GATT) was established following World War II. Early progress within the GATT resulted in the lowering of protection levels for manufactured goods. Trade liberalization for agriculture remained elusive for nearly 50 years. Bringing agriculture into the GATT was difficult for many reasons. One of the reasons for continued protection of agriculture involves the strength of agricultural producer groups and their ability to obtain political support.

Despite the strength and political clout of agricultural producers in DCs, agriculture has been an area of focus in a number of international trade negotiations. The Uruguay Round of the GATT placed restrictions on the levels of agricultural protection that countries can employ. At the same time, numerous regional trade agreements have been made (North American Free Trade Agreement, NAFTA) or are being negotiated (United States–Central America Free Trade Agreement (CAFTA) and Free Trade Area of the Americas (FTAA)) to decrease trade restrictions between countries.

This tendency to liberalize trade is in conflict with the desire of most DCs to provide support to their agricultural producers. While countries struggle with the dilemma of how to liberalize trade and still provide support to their domestic industry, the ultimate solution will likely affect the growth and welfare of both LDCs and DCs. These critical contemporary issues highlight the importance of agricultural trade in our world.

The purpose of this book is to present an analytical framework to critically examine these and other related agricultural trade issues. To accomplish this, an overview of trade theories and topics relevant to the trade of agricultural commodities and products is presented. This textbook is designed for upper-level undergraduate students and entry-level graduate students. It consists of 16 chapters in four parts. Part I focuses on trade theories, including the Ricardian theory, the Heckscher–Ohlin model, and new trade theory. Part II discusses commonly used trade restrictions and agricultural subsidy programs affecting trade flows of agricultural commodities and products from exporting countries to importing countries. Policies included in this part relate to trade protection and some domestic support programs. In addition, this part discusses the implications of multinational (WTO) and regional (NAFTA and FTAA) trade agreements, and the evolution of US trade policy interacting with the trade agreements. Part III concentrates on trade issues related to macroeconomics, including exchange rates and the balance of trade. Part IV focuses on additional trade issues, such as foreign direct investment and multinational firms, economic growth, and the environment.

■ 1.1 THE PURE THEORY OF INTERNATIONAL TRADE

Part I of this text focuses on trade theories, including the Ricardian model, the Heckscher–Ohlin model, and modern trade theory. This part focuses on developing an understanding of the basic economics of trade. The principle of comparative advantage is presented, along with various neoclassical endowment models. In addition, this part exposes the student to theories that account for increasing returns to scale and imperfect competition through new trade theory.

Chapter 2 introduces the concept of the production possibilities frontier and the social indifference curve, under assumptions of the labor theory of value and constant opportunity cost. These concepts are used to present both production and consumption equilibria and gains from trade.

The theory of comparative advantage presented in Chapter 2 is extended in Chapter 3 to the more realistic case of increasing opportunity costs. The restrictive assumption of the labor theory of value is relaxed by introducing two factors of production. The chapter then introduces optimal trade patterns between two countries and the resulting gains from trade. Production and consumption equilibria in a nation are demonstrated in both a closed and an open economy.

Chapter 4 uses endowment theories to show the linkages between factor intensity and factor abundance. The Heckscher–Ohlin theorem and its assumptions are stated. The chapter also provides an overview of the factor-price equalization theorem and its implications for income distribution.

Over time, agriculture has become characterized less as a purely competitive market. As a result, trade theories that account for economies of scale and imperfect competition are more appropriate to analyze agricultural trade than those discussed in the previous parts. Chapter 5 provides an introduction to new trade theory through various topics. Topics discussed in this chapter include the impacts of imperfect competition and economies of scale on competitiveness and international trade. This chapter also introduces the difference between internal and external economies of scale.

■ 1.2 PROTECTION OF THE DOMESTIC INDUSTRY

Agriculture has traditionally received special treatment in countries at all stages of development. Agricultural-related domestic and trade policies have resulted in inefficient agricultural production and distorted trade flows. Part II discusses trade and agricultural policies affecting trade flows of agricultural commodities and products from exporting countries to importing countries. This part includes various types of trade protection and domestic support programs used by exporting and importing countries. In addition, this part discusses the implications of multinational (e.g., World Trade Organization) and regional (e.g., North American Free Trade Agreement and Free Trade Area of the Americas) free trade agreements, and the evolution of US trade policy interacting with the agreements.

Partial equilibrium trade analysis is a useful tool to analyze the welfare effects of agricultural trade policies. Chapter 6 introduces the partial equilibrium trade model through the derivation of the import demand equation and the export supply equation. In this chapter, students learn how an equilibrium is determined in a partial equilibrium framework. This chapter also identifies the divergence in domestic prices due to transportation and other costs.

Given the role of the General Agreement on Tariffs and Trade (GATT) and its successor, the World Trade Organization (WTO), tariffs have been identified as a preferred and transparent means of protection and as a vehicle toward the eventual elimination of trade barriers. Chapter 7 provides an overview of various tariff systems and types of tariffs. Special emphasis is placed on the incidence of tariffs and the determination of the welfare implications of tariffs. The concepts of nominal tariffs and effective protection rates are introduced, while the notion of an optimal tariff is presented.

Chapter 8 provides an overview of both import quotas and export quotas, and their welfare and economic effects. Special emphasis is placed on the comparison of import quotas and tariffs. The tariff rate quota (TRQ), a tool used under various trade agreements to increase market access, is presented. In addition to quotas, various other nontariff trade barriers are discussed in this chapter. The nontariff barriers are international cartels; price discrimination and dumping; export subsidies; and technical and administrative protection.

The Uruguay Round of the GATT was revolutionary in that it was one of the first times that domestic support was included within a multilateral trade agreement. It is important to consider the trade impacts of domestic support policies. To accomplish this, Chapter 9 reviews domestic support mechanisms, including specific production subsidies and variable production subsidies such as deficiency payments and variable levies.

Agriculture was one of the last sectors to be brought fully under the auspices of multilateral trade agreements. Progress has been made toward achieving multilateral trade liberalization, yet much protection remains. Chapter 10 reviews the history of multilateral trade negotiations from the Havana Charter through the various rounds of the GATT and the formation of the WTO under the Uruguay Round, and its potential role in agricultural trade. This chapter also provides the evolution of US trade policy interacting with the multilateral and regional trade negotiations.

While significant progress has been made toward achieving free trade through multilateral trade negotiations, there is a plethora of regional trade agreements throughout the world. Chapter 11 presents the theory of economic integration and identifies the various levels of integration among countries. Students are exposed to the trade diversion and trade creation effects of free trade agreements. The history of various attempts at economic integration is also presented. These include the European Union, NAFTA, and various other free trade areas such as APEC, MERCOSUR, and the potential Free Trade Area of the Americas.

■ 1.3 FOREIGN EXCHANGE MARKETS AND THE BALANCE OF PAYMENTS

Since the early 1970s, fluctuations in currency exchange rates have impacted the flow of agricultural inputs, goods, and services. This part concentrates on trade issues related to macroeconomics, including exchange rates and the balance of trade, and examines their potential impacts on agricultural trade.

An important factor affecting trade flows is the currency exchange rate. Chapter 12 exposes the student to the functioning of foreign exchange markets and their role in determining the exchange rate. The functions of foreign exchange markets are presented and explained. In addition, the determination of currency exchange rates is examined under both the floating and fixed exchange rate systems. Determinants of exchange rates are discussed and their impacts are examined. In addition, this chapter discusses various factors that influence the exchange rate. These factors are examined to show how they influence the equilibrium in foreign exchange markets for the situation in which exchange rates are allowed to float. Additional specifics on the occurrence of the J-curve and derivation of the Marshall–Lerner conditions are included.

Agriculture is a relatively small component of the US economy. As a result, changes in agricultural and agribusiness production, consumption, and trade do not have major impacts on the value of the dollar. The exchange rate can thus be considered as an exogenous variable for US agriculture, as opposed to the situation in developing countries that are more agrarian based. The impact of currency exchange rate fluctuations on agriculture is examined in Chapter 13, using a partial equilibrium and general equilibrium framework. Cases of particular commodities and other historical perspectives are used to show the impact of the exchange rate on agricultural trade.

■ 1.4 DIRECT FOREIGN INVESTMENT, ECONOMIC GROWTH, AND ENVIRONMENT

There are several other trade issues that are not discussed directly in the first three parts, and yet are extremely relevant and timely given the current state of agricultural trade. These other remaining trade issues include foreign direct investment and multinational firms in agriculture and agribusiness; the interdependencies between trade and development; and the relationship between trade and the environment.

Trade in goods and services is only one component of the transfer of resources from one area to another. The movement of capital is also an important dimension to be considered. Chapter 14 introduces the student to the importance of international resource movements such as capital investment by multinational corporations (MNCs). Several real-world cases are used that expose the student to the increasing role of multinational corporations and foreign direct investment (FDI) in the food and agribusiness industries.

Much focus has been placed on the use of trade and trade policies to achieve various welfare objectives. Developed countries are placing an increasing emphasis on the impacts that their agricultural trade policies have on developing countries. Chapter 15 exposes the student to the role of trade in development. Special focus is placed on the increasing role of developing countries in the WTO and their objectives; the willingness of developed countries to make trade concessions to spur economic development in lieu of financial assistance; and the historic role of food aid, through programs such as PL480, and a discussion of their successes and failures.

The Seattle meetings of the WTO emphasized the linkages between international concerns and other issues, namely the environment. Chapter 16 exposes the student to the underlying conflicts between trade and the environment. This is accomplished by comparing the underlying economics of trade and the environment. Students will be exposed to the existence of production externalities and their potential for sub-optimal outcomes. Various issues associated with the lack of harmonization of standards across countries are presented and discussed. The chapter concludes by discussing the environmental consequences of agricultural trade liberalization, as well as the role of environmental issues in achieving multilateral and regional trade agreements.

SELECTED BIBLIOGRAPHY

Dixit, P. and Josling, T. 2001: *The Current WTO Agricultural Negotiations: Options for Progress.* Commissioned Paper No. 18. St. Paul, MN: International Agricultural Trade Research Consortium.

Ingersent, K., Rayner, A. and Hine, R. (eds.) 1994: *Agriculture in the Uruguay Round.* New York: St. Martin's Press.

Josling, T., Tangermann, S. and Warley, T. 1996: *Agriculture in the GATT.* New York: St. Martin's Press.

Sumner, D. 1995: *Agricultural Trade Policy: Letting Markets Work.* Washington, DC: American Enterprise Institute Press.

PART I

Pure Theory of International Trade

Classical Theory of Comparative Advantage

Consumers around the world benefit from trade. Without trade, someone in Anchorage, Alaska, could not enjoy French wines and citizens of France could not eat Alaskan salmon. The factors that determine the pattern of trade between countries can be quite obvious. Greenland has difficulty producing bananas and must import them from elsewhere. But in many instances the factors that influence the pattern of trade are not so clear. In the case of two countries that can both produce wine and cheese, why does one country export wine and the other country export cheese?

To analyze these issues, classical economists adopted the simplifying assumption of the labor theory of value to explain trade patterns between countries. This theory assumes that labor is the only factor of production and that in a closed economy the prices of all commodities are determined by their labor content. The classical theory of international trade is concerned with relative prices as opposed to absolute prices. Given this, goods are exchanged on the basis of the relative amount of labor used to produce goods.

In this chapter, the principles of absolute advantage and comparative advantage are presented. In addition, the production equilibrium that maximizes national income, the consumption equilibrium that maximizes the country's social utility, and the corresponding optimal trade flows are discussed.

■ 2.1 THE PRINCIPLE OF ABSOLUTE ADVANTAGE

Adam Smith pioneered the concept that trade between two countries is based on absolute advantage. If one country produces a commodity more efficiently than another country and is less efficient in producing a second commodity than the other country, then each country can benefit by specializing in the commodity that it produces

Table 2.1 Labor requirements (units) to produce one unit of textiles and one unit of corn

Commodities	USA	UK
Textiles (1 yard)	4	2
Corn (1 ton)	2	4
Total labor	6	6

Table 2.2 Total output (units) before and after trade

	USA	UK	Total output
Before specialization			
Textiles (1 yard)	1	1	2
Corn (1 ton)	1	1	2
After specialization			
Textiles (1 yard)	0	3	3
Corn (1 ton)	3	0	3

more efficiently. Assume that there are two countries, the United States (USA) and the United Kingdom (UK), used to produce corn and textiles. The labor requirements in producing 1 ton of corn are two and four units, respectively, in the USA and the UK. The labor requirements in producing 1 yard of textiles are four and two units in the USA and the UK, respectively. These labor requirements are shown in Table 2.1.

In this particular case, the USA has an absolute advantage in producing corn, because its labor requirement to produce corn is less than that of the UK. Conversely, the UK has an absolute advantage in producing textiles, because fewer units of labor are needed to produce textiles than in the USA. This implies that the USA should specialize in the production of corn and the UK should specialize in the production of textiles, to maximize the total output of these two goods by the two countries. In this example, the USA should export corn to the UK and import textiles from the UK. Similarly, the UK should trade with the USA, exporting textiles to the USA and importing corn from the USA.

If both countries divide their labor between corn and textiles production, the total output of both commodities that they can produce with a given amount of labor (six units in the USA and six units in the UK) is shown in Table 2.2. Prior to specialization, each country can produce 1 ton of corn and 1 yard of textiles. Thus, the total quantity of goods produced by both countries is 2 tons of corn and 2 yards of textiles.

Table 2.2 also shows the total quantities of goods produced by these two countries after they specialize their production on the basis of the principle of absolute advantage and engage in international trade. When the USA specializes in producing corn, it can produce 3 tons of corn with its six units of labor. Similarly, the UK produces 3 yards of textiles with its six units of labor. As a result, international trade increases total output through the specialization of production between countries.

■ 2.2 THE PRINCIPLE OF COMPARATIVE ADVANTAGE

David Ricardo published a book, *Principles of Political Economy and Taxation*, in which he presented the principle of comparative advantage. This is one of the most important trade theories and it has been widely used to analyze trade patterns. Consider an example with labor requirements that are different from those of the previous example. The USA requires four units of labor to produce 1 yard of textiles and two units of labor to produce 1 ton of corn, as shown in Table 2.3. The UK requires six units of labor to produce 1 yard of textiles and 12 units of labor to produce 1 ton of corn.

In this case, the USA has an absolute advantage in producing both corn and textiles. What can we conclude? Should the USA produce both corn and textiles and export them to the UK? The answer is no. The USA cannot trade with the UK if the UK produces nothing. The UK does not have any incentive to trade with the USA, since the UK has nothing to trade with the USA. This implies that the fundamental reason for profitable trade is not in the absolute differences in labor inputs between countries. Instead, the relative labor inputs used to produce commodities are a key determinant of trade.

The USA has an absolute advantage in producing both corn and textiles, because the labor inputs for both commodities are smaller than in the UK. However, the degree of advantage that the USA has over the UK differs with commodities. The USA requires only 1/6 of the labor inputs that the UK requires for corn and 2/3 of the labor inputs that the UK requires for textiles. The absolute advantage for the USA is greater in the production of corn than textiles, indicating that the USA has a comparative advantage in the production of corn.

The UK has an absolute disadvantage in producing both corn and textiles for the same reason that the USA has an absolute advantage in both commodities. However, this disadvantage is smaller in the production of textiles than corn, because the labor required in producing textiles in the UK is 1.5 times the US requirement and that in producing corn is six times greater. This tells us that the UK has a comparative advantage in the production of textiles.

Table 2.3 Labor requirements (units) to produce one unit of commodities X and Y

	USA	UK
Textiles (1 yard)	4	6
Corn (1 ton)	2	12

Table 2.4 Total output before and after specialization

	USA	UK	Total output
Before specialization			
Textiles (1 yard)	1	1	2
Corn (1 ton)	1	1	2
After specialization			
Textiles (1 yard)	0	3	3
Corn (1 ton)	3	0	3

Assume that the USA is endowed with six units of labor and the UK with 18 units of labor. Given the labor endowments, if both countries produce both corn and textiles, each country can produce 1 ton of corn and 1 yard of textiles, as shown in Table 2.4. The total world output is 2 tons of corn and 2 yards of textiles. The total output of corn and textiles increases as the countries specialize on the basis of the principle of comparative advantage. Since the USA has a comparative advantage in producing corn, it specializes in the production of that commodity and can produce 3 tons of corn with its six units of labor. Conversely, the UK specializes in the production of textiles and can produce 3 yards of textiles with its 12 units of labor. Thus, the total world output grows to 3 tons of corn and 3 yards of textiles, indicating that there are increases in total world output through specialization based on the principle of comparative advantage.

■ 2.3 OPPORTUNITY COST AND COMPARATIVE ADVANTAGE

So far, we have assumed that the only factor employed in production is labor in a homogeneous form. This means that there is only one type of labor. But this assumption is not realistic because labor is not homogeneous between two countries, or even within countries. Gottfried Haberler used the concept of "opportunity cost" to explain the theory of comparative advantage, without introducing the restrictive assumption of homogeneous labor. The opportunity cost is defined as the minimum amount of a second commodity that must be given up to produce one additional unit of the first commodity.

The opportunity cost of textiles in terms of corn in the USA is defined as the minimum amount of corn that the USA has to give up to produce an additional unit of textiles. Similarly, the opportunity cost of corn in terms of textiles in the USA is defined as the minimum amount of textiles that the USA has to give up to produce an additional unit of corn. Drawing from the previous example in Table 2.3, the opportunity costs are calculated in Table 2.5.

The USA requires four units of labor to produce 1 yard of textiles and two units of labor to produce 1 ton of corn. If the USA wants to produce more corn, it can produce 2 tons of corn for every yard of textiles it gives up. The opportunity cost

Table 2.5 Opportunity costs (units) in producing commodities X and Y

	USA	UK
Textiles (1 yard)	2	1/2
Corn (1 ton)	1/2	2

of 1 yard of textiles is 2 tons of corn. In other words, 2 tons of corn must be given up in order to produce one additional yard of textiles. Similarly, if the USA wants to produce one additional ton of corn instead of textiles, it has to give up only 1/2 yard of textiles to produce one additional ton of corn. The opportunity cost of producing 1 ton of corn is 1/2 yard of textiles. Using the same method, we can calculate opportunity costs for the UK. The opportunity cost of textiles is 1/2 ton of corn in the UK and that of corn is 2 yards of textiles.

The USA has a comparative advantage in the production of corn, because the opportunity cost of corn in terms of textiles is lower in the USA than in the UK (Table 2.5). Conversely, the UK has a comparative advantage in the production of textiles, because the UK has a lower opportunity cost of textiles than the USA. This implies that, to maximize outputs of corn and textiles, the USA should specialize in the production of corn and the UK in the production of textiles. As shown in section 2.2, when the two countries specialize their production, the total world output is 3 yards of textiles and 3 tons of corn.

BOX 2.1 COMPARATIVE ADVANTAGE IN AGRICULTURE: CENTRAL AMERICA AND THE CARIBBEAN VERSUS THE USA

Countries in the Western Hemisphere are characterized by a wide range of differences in resource endowments and economic conditions. This region includes some of the very richest and some of the very poorest countries in the world. A number of countries have a tropical climate, while others are in the temperate zone. Other differences exist concerning investment, technology, education, and so on. These differences are pronounced in the agricultural sector. Countries in Central America and the Caribbean (CA&C) have a tropical climate. As such, it would be expected that they have an advantage in the production of tropical agricultural products. The USA, with a majority of its land located in the temperate region, would likely have an advantage in the production of commodities that require a shorter growing season.

Two commodities that provide a good example of comparative advantage are bananas and corn. The CA&C region has traditionally been a banana exporter, while the USA is a dominant exporter of corn. Is this the result of absolute or comparative advantage? Tables 2.6 and 2.7 provide data for banana and corn yield, production, and trade for the CA&C region and the USA. The USA has a

Table 2.6 Banana yield, production, and trade for the USA and Central America and the Carribean, 2000

	CA&C	USA
Yield (Hg/ha)	210,093	222,572
Production (Mt)	8,304,729	13,154
Net exports (Mt)	4,116,417	−3,630,448

Source: FAOSTAT (2003), United Nations, Rome

Table 2.7 Corn yield, production, and trade for the USA and Central America and the Carribean, 2000

	CA&C	USA
Yield (Hg/ha)	22,631	85,910
Production (Mt)	20,724,345	251,854,000
Net exports (Mt)	−8,232,610	47,677,560

Source: FAOSTAT (2003), United Nations, Rome

clear advantage in the production of corn, with a yield nearly four times greater than that of CA&C. What may be surprising is that, for this time period, the USA has a slight yield advantage in producing bananas over the CA&C region.

This data would indicate that, if absolute advantage determined trade flows, the USA should export both commodities.[a] But the USA exports corn and imports bananas. The USA has an absolute advantage in producing both corn and bananas, because it has higher yields in producing both commodities. However, the USA has a greater advantage in producing corn than bananas, indicating that it has a comparative advantage in producing corn over CA&C. On the other hand, CA&C has an absolute disadvantage in producing corn and bananas compared to the USA. However, their disadvantage is smaller in producing bananas than corn, indicating that CA&C has a comparative advantage in producing bananas over the USA. This is consistent with the data showing the USA as a net exporter of corn and a net importer of bananas, while the CA&C region is a net exporter of bananas and a net importer of corn. Although the USA has an absolute advantage in the production of both commodities, the magnitude of its advantage in corn production is much greater than in banana production, giving the USA a comparative advantage in corn production and the CA&C region a comparative advantage in banana production.

[a] Note that not all parts of the USA can produce bananas. Despite the yield advantage for this particular year, it is not likely that the USA has the capacity to meet its own demand for bananas.

■ 2.4 THE TERMS OF TRADE AND THE TRADE PATTERN

Given the data from the previous example in Table 2.3, the pre-trade price ratio in each of the two countries can be determined under an assumption of the labor theory of value, as follows:

$$\left(\frac{P_t}{P_c}\right)_{us} = \left(\frac{a_t}{a_c}\right)_{us} = \frac{4}{2} = 2 \tag{2.1}$$

$$\left(\frac{P_t}{P_c}\right)_{uk} = \left(\frac{b_t}{b_c}\right)_{uk} = \frac{6}{12} = 0.5 \tag{2.2}$$

where P_t and P_c are the price of textiles per yard and the price of corn per ton, respectively, a_t and a_c are the units of labor used to produce 1 yard of textiles and 1 ton of corn, respectively, in the USA, and b_t and b_c are those required in the UK. The price ratios in equations (2.1) and (2.2) represent the price of textiles in terms of corn. In the USA, 1 yard of textiles can be exchanged for 2 tons of corn. Similarly, in the UK, 1 yard of textiles can be exchanged for 0.5 tons of corn.

International trade occurs because of differences in the price ratio of textiles to corn between the two countries. If the prices of the commodities are the same between the two countries, the commodities are not traded. Since the relative price of textiles in the USA (2 tons of corn) is higher than that in the UK (0.5 tons of corn), the USA starts to import textiles from the UK. As a result, the price of textiles declines in the USA because its import of textiles increases the total supply of textiles, and the price increases in the UK because its export of textiles increases the total demand for textiles. On the other hand, the UK starts to import corn from the USA, since the price of the corn in the USA (0.5 yards of textiles) is lower than in the UK (2 yards of textiles). This will lower the price of corn in the UK and raise it in the USA. The two countries continue to trade the goods with each other until the price ratio in the USA equals the price ratio in the UK. At that price ratio, US exports of corn equal UK imports of corn, while US imports of textiles are the same as UK exports of textiles. This price ratio is known as the terms of trade.

Specifically, the terms of trade for a nation is defined as the ratio of the price of its export commodity to the price of its import commodity. The UK's terms of trade in the previous example is P_t/P_c, while the USA's terms of trade is P_c/P_t. In the case in which many commodities are traded between two countries, the terms of trade of a nation is measured by the ratio of the price index of its exported goods to the price index of its imported goods. The ratio is generally multiplied by 100 in order to express the terms of trade as a percentage. Country A's terms of trade are calculated as

$$\text{TOT}_a = P_x/P_m \tag{2.3}$$

where P_x is the price of an export commodity or the index of its export prices and P_m is the price of an import commodity or the index of its import prices. The terms

of trade of country B, which is a trading partner of country A, is the inverse or reciprocal of country A's terms of trade, because A's exports are equal to B's imports,
and B's exports are the same as A's imports.

In the above example, the terms of trade of the UK will be determined somewhere
between the two countries' pre-trade price ratios. Its terms of trade could be 1.5,
indicating that the UK exports 1.0 yards of textiles in exchange for 1.5 tons of corn
produced in the USA. Then, the USA's terms of trade would be 0.67 (1/1.5), indicating that the USA exports 1 ton of corn in exchange for 0.67 yards of textiles produced in the UK. If the UK's terms of trade changes from 1.5 to 1.0, this would indicate
that the UK's price of textiles falls by 33%, or the US export price rises by 33%,
indicating that the UK exports 1 yard of textiles in exchange for 1 ton of corn produced in the USA.

■ 2.5 PRODUCTION POSSIBILITIES FRONTIER AND CONSTANT OPPORTUNITY COST

The production possibilities frontier (PPF) is a curve that shows the alternative combinations of two commodities that a country can produce by fully utilizing its resources.
We can derive the PPF under the following assumptions:

1 The USA is endowed with L_a units of labor and the UK with L_b units of labor.
2 The production of corn and textiles requires a_c and a_t units of labor per unit of
   output, respectively, in the USA and b_c and b_t units of labor, respectively, in the
   UK.

In the equilibrium, the total units of labor available in a country are equal to the
total units of labor employed to produce corn and textiles in that country. This is
expressed for the USA as

$$L_a = a_t Q_t + a_c Q_c \tag{2.4}$$

where Q_t and Q_c represent the quantities of textiles and corn produced, respectively,
and L_a is the total units of labor available in the USA.

The USA maximizes production of corn and textiles under an equilibrium condition in the labor market, as shown in equation (2.4). Given equation (2.4), the economy's total labor force is fully employed. If the USA produces only textiles, it can
produce L_a/a_t yards of textiles. Similarly, the USA can produce L_a/a_c tons of corn if
it produces only corn. These are extreme cases. If the USA decides to produce both
corn and textiles, there are many possible combinations of quantities of corn and
textiles that can be produced. The quantity combinations of corn and textiles are
expressed on a straight line with a downward slope. In Figure 2.1, the y-axis interception point, L_a/a_c, shows the maximum quantity of corn that the USA can produce.
The x-axis interception point, L_a/a_t, shows the maximum quantity of textiles that the
USA can produce. The straight line connecting the interception points on the x- and

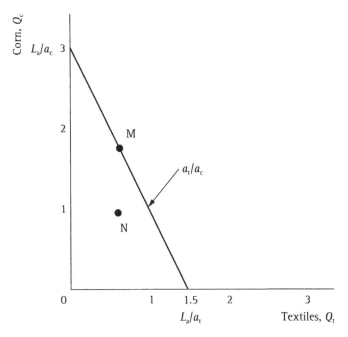

Figure 2.1 The US production possibilities frontier.

y-axes (L_a/a_c on the y-axis and L_a/a_t on the x-axis) is known as the production possibilities frontier (PPF).

Assume that the labor endowment in the USA is six units. With the given labor requirements for producing corn and textiles in the USA in Table 2.3, equation (2.4) can be written as

$$6 = 4Q_t + 2Q_c \tag{2.5}$$

The maximum amount of textiles that the USA can produce is 1.5 yards of textiles ($L_a/a_t = 6/4$) and the maximum amount of corn production is 3 tons of corn ($L_a/a_c = 6/2$). These are the interception points on the x-axis and the y-axis, respectively, in Figure 2.1. The PPF of the USA is a straight line connecting the two interception points. Any point in the PPF shows the feasible combinations of textiles and corn that the USA can produce under the given labor constraints. Any point on the PPF, such as point M, implies that the economy is making full use of its resources in producing the commodities and satisfies the labor market equilibrium condition in equation (2.5). Any point inside the PPF, such as point N, implies that the economy is not making full use of its resources in producing these commodities. Any point outside the PPF is unattainable given the country's resources. The slope of the PPF is

$$\left(\frac{L_a/a_c}{L_a/a_t} = \frac{a_t}{a_c} \right)$$

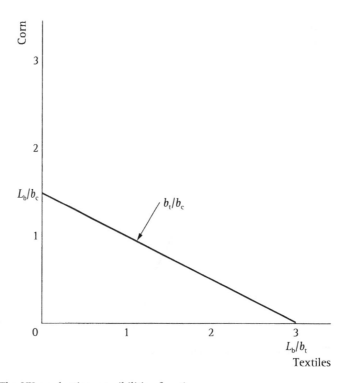

Figure 2.2 The UK production possibilities frontier.

The slope (a_t/a_c) represents the opportunity cost of textiles in terms of corn, because the slope represents the quantity of corn that the economy must give up to add one unit of textiles. In the above example, the opportunity cost of textiles in terms of corn is 2 $(a_t/a_c = 4/2)$, which is the slope of the PPF in the USA. Since the PPF is linear, the opportunity cost is constant, indicating that the opportunity cost of textiles in terms of corn remains the same at any point on the PPF.

The PPF in the UK can be obtained in a similar way. The total units of labor available in the UK must be equal to the total units of labor employed in the country. This is expressed as follows:

$$L_b = b_t Q_t + b_c Q_c \tag{2.6}$$

This equation shows the labor market equilibrium condition for the UK. When this condition holds, the total labor force in the UK is fully employed.

The PPF curve obtained from equation (2.6) is shown in Figure 2.2. The maximum amount of corn that the UK can produce is L_b/b_c and the maximum amount of textiles is L_b/b_t. The absolute slope of the PPF (b_t/b_c) represents the opportunity cost of textiles in terms of corn.

Assume that the total amount of labor endowment in the UK is 18 units. With the given labor requirements from Table 2.3, equation (2.6) can be rewritten as

$$18 = 6Q_t + 12Q_c \tag{2.7}$$

The maximum amount of corn that the UK can produce is 1.5 tons ($L_b/b_c = 18/12$) and maximum amount of textiles is 3 yards ($L_b/b_t = 18/6$). The PPF of the UK is a straight line connecting the two interception points, as shown in Figure 2.2. The slope of the PPF, the opportunity cost of textiles in terms of corn, is 0.5 ($b_t/b_c = 6/12$).

The USA has a comparative advantage over the UK in producing textiles if the US opportunity cost in producing textiles (a_t/a_c) is lower than the UK opportunity cost (b_t/b_c). Conversely, the USA has a comparative advantage over the UK in producing corn if the US opportunity cost of corn is lower than the UK opportunity cost. In this example, the opportunity cost of textiles in terms of corn is 2 in the USA and 0.5 in the UK, indicating that the UK has a comparative advantage over the USA in producing textiles. Conversely, the opportunity cost of corn in terms of textiles is 0.5 in the USA and 2 in the UK. Therefore, the USA has a comparative advantage over the UK in producing corn.

■ 2.6 PRODUCTION EQUILIBRIUM AND THE CONSUMPTION POSSIBILITIES FRONTIER

Assume that the prices of textiles and corn in the USA are P_t and P_c, respectively. Then the value of total output (V) produced in the USA is

$$V = P_t Q_t + P_c Q_c \tag{2.8}$$

where Q_t is the total quantity of textiles and Q_c is the total quantity of corn produced by the USA. In a simple economy, since the total value of output in a nation represents its total income, equation (2.8) also represents a nation's income. This equation shows the amount of textiles and corn that the USA must produce to generate the national income of V, given output prices. For the given level of V, the relationship between textiles and corn is linear in equation (2.8). If the USA specializes in producing textiles, it will produce V/P_t units to generate its income of V. On the other hand, if the USA specializes in producing corn, it will produce V/P_c units of corn. V/P_t and V/P_c are the x-axis and y-axis interception points, respectively, in Figure 2.3. A straight line connecting the two interception points in Figure 2.3 represents income for the USA. The slope of the income line is a price ratio of textiles to corn, as

$$\frac{V/P_c}{V/P_t} = \frac{P_t}{P_c} \tag{2.9}$$

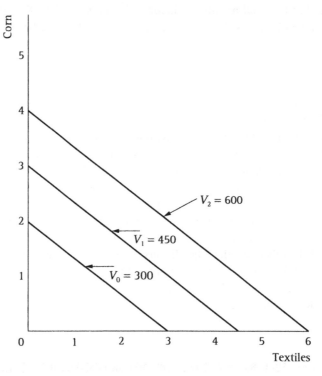

Figure 2.3 A family of income lines.

Assume that the price of corn is $150 per ton and that of textiles is $100 per yard. Equation (2.8) can be rewritten as

$$V = 100Q_t + 150Q_c \qquad\qquad (2.10)$$

This equation shows alternative combinations of textiles and corn that the USA must produce to generate a certain level of national income. For instance, to obtain an income level of $300, the USA could produce either 3 yards of textiles ($300/100) or 2 tons of corn ($300/150). Alternatively, the USA could produce some combination of both products on the line V_0 in Figure 2.3. The line V_0 represents alternative combinations of textiles and corn that the USA can produce to generate an income of $300.

Similarly, to obtain an income equivalent to $450, the USA could produce 3 tons of corn ($450/150) or 4.5 yards of textiles ($450/100), or produce some combination of both products on the line V_1. The line V_1 shows alternative combinations of textiles and corn that the USA should produce to generate an income of $450. The line V_2 also represents alternative combinations of textiles and corn that the USA should produce to generate an income of $600.

In Figure 2.3, income lines such as V_1 and V_2, which are above V_0, represent higher levels of income than that represented by V_0. Also, income line V_2 represents a higher level of income than income line V_1. US income increases when its income line moves away from the origin.

The US economy seeks to maximize its national income, given its production capacity of textiles and corn represented by its PPF. This can be expressed in the following way:

Max $V = P_t T + P_c C$
subject to
$L = a_t T + a_c C$ (2.11)

The quantities of textiles and corn that maximize the economy's total income under the given labor market equilibrium condition can be found by presenting income lines and the PPF in the same two-dimensional graph. In Figure 2.4, the straight line KM is the PPF for the USA. Its slope (a_t/a_c) is −2, based on the previous example in Table 2.3. The straight line DF is the economy's income line with the given prices of textiles ($100) and corn ($150). The slope of line DF (P_t/P_c) is −0.67 in Figure 2.4. The opportunity cost of textiles in the USA (2 tons of corn) is higher than the relative price of textiles in the world economy (0.67), indicating that the USA has a

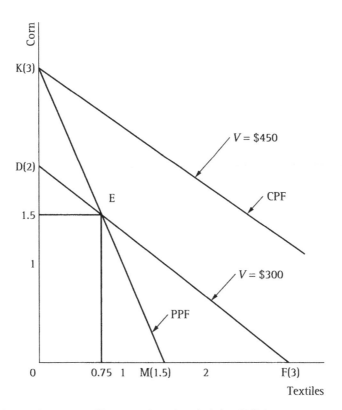

Figure 2.4 The production equilibrium point when $(a_t/a_c) > (P_t/P_c)$.

comparative disadvantage in producing textiles. This implies that the USA has a comparative advantage in producing corn.

The economy must produce somewhere on the PPF. Given the fixed levels of resources, it is impossible for the USA to produce outside the PPF. On the other hand, the USA does not want to produce at a point inside the PPF, because this does not satisfy the equilibrium condition in the labor market.

Assume that the economy produces at a point on the PPF (KM), such as E. Next, draw the income line of the USA associated with the production point E (DF). A straight line, DF, represents US income associated with the production point, E. The income line shows the maximum amount of textiles and corn that the economy must produce to generate a given income represented by DF. The economy produces 1.5 tons of corn and 0.75 yards of textiles, and generates an income of $300 ($150 \times 1.5 + 100 \times 0.75$).

As the production point slides down on the PPF (from E to M), the income line shifts inward, indicating that the USA's income decreases. Alternatively, as the production point slides upward on the PPF (from E to K), the income line shifts outward, indicating that US income is increasing. When the production point coincides with K, the income line reaches its highest level. This means that the economy maximizes its national income when it produces at point K. The economy's income is $450 by producing 3 tons of corn (150×3). Therefore, point K is the production equilibrium of the USA. The income line, which corresponds to production equilibrium point K, is the consumption possibilities frontier (CPF) because the economy's income can be spent for these two commodities. Consumers can consume at any point on the CPF. Note that the CPF lies totally outside the PPF except at the single point K. This means that free trade makes it possible to consume beyond the boundary of PPF.

Now let us assume that the slope of the income line (P_t/P_c) is steeper than the PPF (a_t/a_c), as shown in Figure 2.5. For example, assume that P_c and P_t are $100 per ton and $250 per yard. Then the income equation can be written as $V = 250Q_t + 100Q_c$. In this figure, the straight line KM is the USA's PPF and a line FD, passing through point E on the PPF, is the economy's income line. When the economy produces at point E (1.5 tons of corn and 0.75 yard of textiles), the economy's income is $337.5 ($250 \times 0.75 + 150 \times 1.5$). The USA achieves a higher income level as the production point slides down the PPF. The USA can maximize its income by moving its production from E to M. It specializes in the production of textiles to maximize its income at $375. The production point M, therefore, is the production equilibrium for the USA. The line MN passing through the production equilibrium point, M, is the CPF for the USA. The USA produces OM units of textiles and consumes both commodities at any point on the CPF through trade. The CPF lies beyond the economy's PPF except at the single point M.

If the slope of the PPF is equal to the slope of the income line, the CPF coincides with the PPF. Production can take place at any point on the PPF. In this case, the economy will gain nothing from international trade.

As shown in Figures 2.4 and 2.5, the USA specializes in producing one commodity if its opportunity cost is lower than the terms of trade for that commodity. This

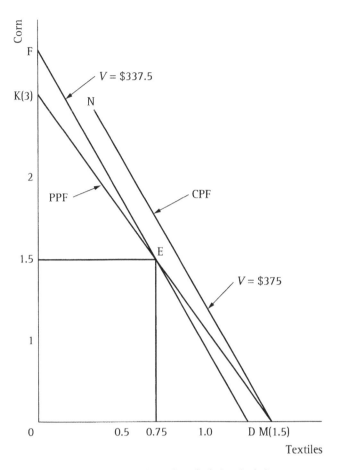

Figure 2.5 The production equilibrium point when $(a_t/a_c) < (P_t/P_c)$.

indicates that an economy will specialize completely in producing one commodity when there are constant opportunity costs.

■ 2.7 CONSUMPTION EQUILIBRIUM AND THE SOCIAL INDIFFERENCE CURVE

To determine the consumption point on the CPF, let us introduce the economy's social indifference map. A social indifference curve (SIC) for an economy is defined as all combinations of goods that the economy can consume to obtain the same level of social utility. The curve is convex from the origin in a two-dimensional graph. For example, social indifference curve SIC_1 in Figure 2.6 shows alternative combinations of corn and textiles that an economy can consume to achieve the same level of social utility achieved at point E_1 (SIC_1). There are an infinite number of social indifference curves in the two-dimensional graph, and they do not intersect with one another.

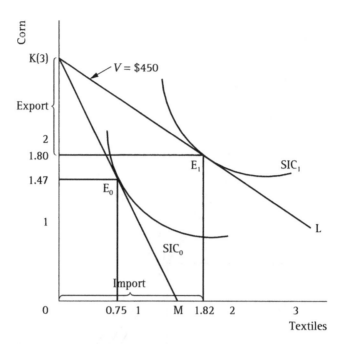

Figure 2.6 Production and consumption equilibrium points in the USA.

The level of social utility increases as the social indifference curves move farther away from the origin. A country tries to reach the highest social indifference curve given its consumption possibilities frontier.

In Figure 2.6, a straight line, KM, is the USA's PPF. The slope of the PPF is -2, as in the previous example. Another linear line, KL, is the USA's CPF, or income line, associated with the equilibrium production point K. As discussed in the previous section, the USA can maximize its income when it produces 3 tons of corn with the international price ratio ($P_t/P_c = 0.67$) lower than its opportunity cost ($a_t/a_c = 2$). The equilibrium before trade occurs at point E_0, where the PPF is tangent to the highest attainable social indifference curve, SIC_0. In other words, before trade, the USA produces and consumes at point E_0, indicating that 1.47 tons of corn and 0.75 yards of textiles are produced and consumed. After opening trade, production occurs at point K and consumption occurs at point E_1, where the CPF is tangent to the social indifference curve SIC_1. This is the highest social indifference curve that the USA can reach. It can maximize its utility by consuming 1.8 tons of corn and 1.82 yards of textiles. Since the USA specializes in the production of corn, producing 3 tons of corn, it imports 1.82 yards of textiles and exports 1.2 tons of corn to maximize its welfare. A movement from social indifference curve SIC_0 to SIC_1 is an increase in social utility from international trade and is an example of the gains from trade.

A similar analysis can be done for the UK with the example given in the previous section. In Figure 2.7, a straight line, KM, is the UK's PPF and another straight line, LM, is its CPF, or income line, associated with production point M. The UK specializes

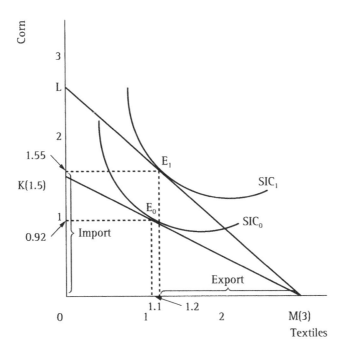

Figure 2.7 Production and consumption equilibrium points in the UK.

in producing textiles with the international price ratio greater than its opportunity cost. The UK maximizes its income when it produces 3 yards of textiles. The UK's equilibrium before trade occurs at point E_0, indicating that it produces and consumes 0.92 tons of corn and 1.1 yards of textiles. After opening trade with the USA, production occurs at point M and consumption occurs at point E_1, where the CPF is tangent to the social indifference curve SIC_1, the highest attainable social indifference curve that the UK can reach. The UK can maximize its utility by consuming 1.55 tons of corn and 1.2 yards of textiles. Since the UK specializes in producing 3 yards of textiles, the UK imports 1.55 tons of corn from the USA and exports 1.8 yards of textiles to the USA.

The equilibrium terms of trade is determined when US exports of corn are equal to UK imports of corn and UK exports of textiles equals US imports of textiles. The determination of the terms of trade will be discussed further in Chapter 3.

SUMMARY

1 Adam Smith argued that countries specialize in the production of commodities on the basis of absolute advantage, and exchange part of their output for commodities produced in other countries. Each country can produce and consume more, indicating that trade is mutually beneficial. However, the principle of absolute advantage cannot be generalized to explain all trade between nations.

2   David Ricardo introduced the principle of comparative advantage. He argued that even though one country has absolute advantage in the production of all commodities, the country should specialize in producing commodities in which it has greater advantage. The other country should specialize in producing the commodities in which it has a smaller disadvantage. In this case, both countries will produce and consume more by specializing in the production of one commodity and exchanging their output.

3   Gottfried Haberler used the concept of opportunity cost to explain the principle of comparative advantage. The opportunity cost of a commodity is defined as the amount of the other commodity that one must give up to produce an additional unit of the commodity. Haberler stated that a country has a comparative advantage in the production of a commodity over the other country if its opportunity cost of the commodity is lower than that of the other country.

4   A country's PPF can be derived from its production function and the equilibrium condition in its labor market. The PPF shows alternative combinations of two commodities that the country can produce by fully utilizing its resources (labor). The country's objective is to maximize the total value of output under the constraint of the PPF. The production equilibrium is obtained at the point at which the PPF intersects the highest income line. An income line associated with the production equilibrium point is known as the consumption possibilities frontier (CPF), implying that consumers in the country can consume at any point on the CPF.

5   The consumption equilibrium is obtained at the point at which the CPF is tangent to the highest attainable social indifference curve. This indicates that consumers can maximize their satisfaction at the consumption point. Since the consumption equilibrium point differs from the production equilibrium point, the country exports surplus production and imports the other commodity.

6   The gain from trade is the outward movement of the social indifference curve from the lower level to the upper level.

7   Under an assumption of constant opportunity costs (linear PPF) in a two-country world, one country specializes in producing one commodity while the other country specializes in producing the other good, indicating that the two countries establish complete specialization.

KEY CONCEPTS

Consumption equilibrium – Quantities of two goods consumed to maximize consumers' utility.

Consumption possibilities frontier – An income line corresponding to the production equilibrium point.

Gains from trade – An increase in social welfare through international trade.

Income line – A curve representing alternative quantities of two goods to be produced that generate a certain level of national income.

Opportunity cost – Cost of production of a good in term of alternative goods.

Pattern of trade – The direction of goods traded between two countries.

Principle of absolute advantage – Export opportunity based on lower production cost.

Principle of comparative advantage – Export opportunity based on lower production cost relative to other goods.

Production equilibrium – Quantities of two goods produced to maximize producers' income.

Production possibilities frontier – A curve representing alternative quantities of two goods that can be produced with given resources.

Social indifference curve – A curve representing all combinations of goods that the economy can consume to obtain the same level of social utility.

Terms of trade – The ratio of the price of export commodities to the price of import commodities.

QUESTIONS AND TASKS FOR REVIEW

1 According to Adam Smith, what is the basis for trade? How are gains from trade generated?
2 Does the principle of absolute advantage explain all trade?
3 Explain the principle of comparative advantage and differences between it and the principle of absolute advantage.
4 What is the terms of trade and how is it obtained?
5 Explain the principle of comparative advantage using the concept of opportunity cost.
6 What is the production possibilities frontier (PPF)?
7 What is the consumption possibilities frontier (CPF) and why is it called the CPF?
8 How can the production equilibrium be defined?
9 How can the consumption equilibrium be defined?

SELECTED BIBLIOGRAPHY

Bhagwati, J. N. 1964: The pure theory of international trade: a survey. *Economic Journal*, 74, 1–84.

Haberler, G. 1936: *The Theory of International Trade with its Applications to Commercial Policy*, trans. A. Stonier and F. Benham. London: Hodge. See Chs. 9 and 10.

Ricardo, D. 1911: *The Principles of Political Economy and Taxation*. London: Dent/New York: Dutton. See Ch. 7.

Smith, A. 1806: *An Inquiry into the Nature and Causes of the Wealth of Nations*. Edinburgh and London. See Chs. 1 and 3.

Comparative Advantage with Two Factors of Production

■ 3.0 INTRODUCTION

The previous chapter analyzed the pattern of trade between countries under the assumption that labor is the only factor of production (the labor theory of value). However, the assumption is quite simplistic and restrictive. The real world utilizes many types of input, including capital and land, to produce goods and services. Countries are endowed with different types and quantities of these resources. Given the available production technology, a country's endowments influence its production efficiency.

   This chapter extends the theory of comparative advantage to the more realistic case of increasing opportunity costs. In addition, the restrictive assumption of the labor theory of value is relaxed by introducing two factors of production. The concept of optimal trade patterns between two countries and the resulting gains from trade is then introduced. Production and consumption equilibria in a nation are demonstrated for the case of both a closed and an open economy.

■ 3.1 A PRODUCTION POSSIBILITIES FRONTIER WITH INCREASING OPPORTUNITY COSTS

Consider an economy that produces two commodities X and Y, by using two input factors, capital (K) and labor (L). Production functions for commodities X and Y are

$$Q_x = f_x(K_x, L_x) \tag{3.1}$$

$$Q_y = f_y(K_y, L_y) \tag{3.2}$$

where Q_x is the quantity of commodity X produced, Q_y is the quantity of commodity Y produced, K_x and K_y are the quantities of capital used to produce commodities X and Y, respectively, and L_x and L_y are the quantities of labor used to produce commodities X and Y.

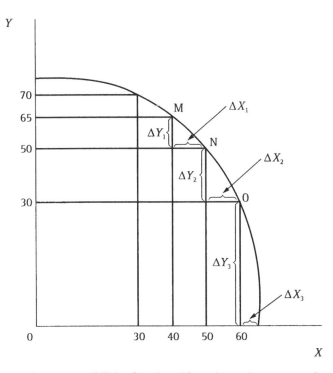

Figure 3.1 The production possibilities frontier with an increasing opportunity cost of commodity X in terms of commodity Y.

Equilibrium conditions in capital and labor markets in an economy are

$$K = K_x + K_y \tag{3.3}$$

$$L = L_x + L_y \tag{3.4}$$

where K and L are the total capital and labor available in the economy. Equation (3.3) is the economy's equilibrium condition in the capital market, and equation (3.4) represents the economy's equilibrium condition in the labor market. We can derive the economy's production possibilities frontier (PPF), which shows the alternative combinations of commodities X and Y that this economy can produce under equilibrium conditions in both capital and labor markets (equations (3.3) and (3.4)). As shown in Figure 3.1, the economy's PPF is assumed to be concave from the origin, indicating that these capital and labor resources are not equally used in the production of commodities X and Y.

The PPF shows the maximum amount of commodities X and Y that this economy can produce with its given resources. When the economy produces commodities X and Y along the PPF, the economy is in equilibrium in its capital and labor markets.

When the production point moves from M to N, the economy must give up ΔY_1 units of Y (15 units) to get an additional ΔX_1 units of X (10 units). Hence, the ratio

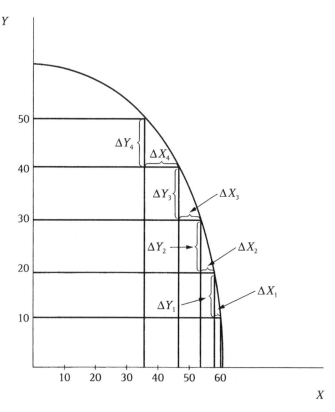

Figure 3.2 The production possibilities frontier with an increasing opportunity cost of commodity Y in terms of commodity X.

$\Delta Y/\Delta X$ represents the opportunity cost of X in terms of Y. This ratio is also an absolute slope of the PPF. Since the PPF is concave from the origin, the opportunity cost of X in terms of Y increases as more X is produced. As the production point on the PPF moves from N to R, the economy gives up ΔY_2 units of Y (20 units) to produce an additional ΔX_2 units of X (10 units). This implies that the opportunity cost of X in terms of Y at the production point R is larger than that at production point N, indicating an increasing opportunity cost. The opportunity cost of X in terms of Y is also known as the marginal rate of transformation (MRT) of X for Y, defined as the amount of Y the economy is willing to trade in exchange for an amount of X.

Figure 3.2 exhibits an increasing opportunity cost of commodity Y in terms of X. For each additional 10 units of Y that the country produces, it must give up more and more units of X (ΔX_3 is larger than ΔX_2 and ΔX_2 is larger than ΔX_1), indicating increasing opportunity costs of Y in terms of X.

The allocation of resources such as labor and capital in producing X and Y is optimal at any point on the PPF, because the curve represents the maximum quantity of commodities X and Y that the economy can produce with its given resources.

3.2 EQUILIBRIUM IN A CLOSED ECONOMY

Figure 3.3 portrays the equilibrium in a closed economy. Curve KM is the PPF of the economy. It is concave to the origin and exhibits increasing opportunity costs in producing X and Y. The equilibrium occurs at point E, where the PPF is tangent to the highest attainable social indifference curve (SIC) in a closed economy (autarky). At point E, the economy can maximize its social utility under the production constraint given by its PPF, since SIC_2 is the highest social indifference curve that the economy can reach under the constraints of the PPF.

The economy produces and consumes 40 units of commodity X and 40 units of commodity Y. This country neither exports nor imports. The common absolute slope of the PPF and SIC_2 at point E represents the pre-trade equilibrium price ratio, P_x/P_y, or the autarky price ratio.

The line represented by the price ratio is known as the income line, and it reflects the economy's total value of outputs at the given prices of commodities X and Y. At point E, the price ratio (P_x/P_y) is equal to the opportunity cost of X in terms of Y.

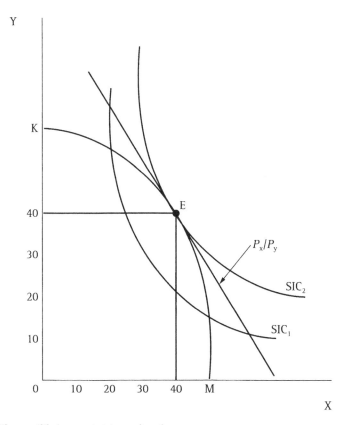

Figure 3.3 The equilibrium point in a closed economy.

3.3 PRODUCTION EQUILIBRIUM IN AN OPEN ECONOMY

Assume that the price of X is P_x and the price of Y is P_y. The total value of output produced by the economy at these prices is

$$V = P_x Q_x + P_y Q_y \tag{3.5}$$

This equation shows combinations of commodities X and Y that the economy can produce to generate the given income, V. As discussed in the previous chapter, an income line is derived from equation (3.5). The income line represents the income generated in the economy by producing commodities X and Y. The slope of the income line is the terms of trade $-(P_x/P_y)$ in an open economy.

As the income line shifts outward (inward), national income increases (decreases). Each income line shows the alternative combinations of commodities X and Y that the economy can produce to generate the economy's national income corresponding to the income line. All income lines have the same slope, represented by $-P_x/P_y$. The only difference is that each income line represents different levels of income. In Figure 3.4, each income line (V_0 through V_4) represents different levels of the economy's income; a higher income line indicates a higher level of income.

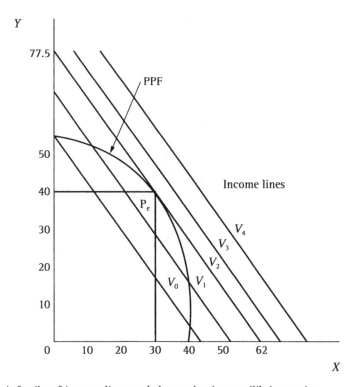

Figure 3.4 A family of income lines and the production equilibrium point.

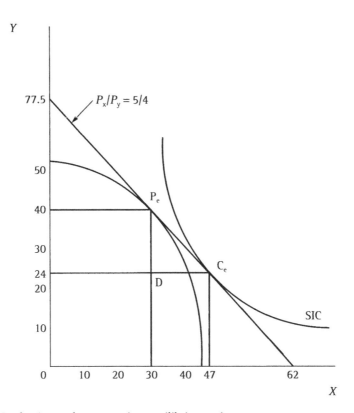

Figure 3.5 Production and consumption equilibrium points.

The economy seeks to maximize national income, given the constraints of the PPF. In Figure 3.4, national income is maximized when the economy produces at point P_e, where the PPF is tangent to income line V_2. Production of commodities X and Y at any point other than point P_e gives a national income smaller than V_2. Production point P_e is, therefore, the economy's equilibrium production point. The economy cannot reach a higher income line than V_2 because of the constraint imposed by the PPF. The income line V_2, tangent to the PPF, is called the consumption possibilities frontier (CPF) because the economy's income, represented by the income line, can be spent for commodities X and Y. Consumers in the economy can consume at any point on the CPF.

The economy produces 30 units of commodity X and 40 units of commodity Y at production point P_e in Figure 3.5. The economy can consume at any point on the CPF. Observe that the CPF lies totally outside the PPF, except at the single point P_e. This means that free trade increases the economy's social utility with an appropriate distribution of income after trade.

■ **3.4 CONSUMPTION EQUILIBRIUM AND THE PATTERN OF TRADE**

Once the CPF has been determined, the next step is to determine where on the CPF the economy should consume. As discussed in the previous section, the economy can maximize its income when it produces at production point P_e (40 units of Y and 30 units of X). Assume that the price of X (P_x) is $5 and the price of Y (P_y) is $4. The total value of output produced by the economy is $310 ($5 × 30 + $4 × 40) at production point P_e. A line tangent to the PPF at production point P_e is the CPF in Figure 3.5. The slope of the CPF is the terms of trade, which is equal to 1.25 (5/4). The CPF represents an income of $310. The consumption equilibrium occurs at point C_e where the CPF is tangent to the highest attainable SIC. Since this social indifference curve is the highest one that the economy can reach, the economy's social utility is at its maximum when the economy consumes at point C_e.

With the production and consumption equilibrium points (P_e, C_e), the economy produces 30 units of commodity X and 40 units of commodity Y, and consumes 47 units of commodity X and 24 units of commodity Y. Given this equilibrium, the economy exports 16 units of commodity Y and imports 17 units of commodity X. As shown in Figure 3.5, this country is not completely specialized in one commodity. This occurs because of increasing opportunity costs.

Observe that the production and consumption points are on the same CPF. This implies that the value of goods produced is equal to the value of goods consumed. The triangle $P_e DC_e$ is known as the trade triangle, because it represents the quantities of commodities that the country exports and imports to simultaneously maximize producers' profit and consumers' social utility for a specific terms of trade.

BOX 3.1 INCOMPLETE SPECIALIZATION: THE CASE OF THE USA–MEXICO VEGETABLE TRADE

The North American Free Trade Agreement (NAFTA) came into effect on January 1, 1994. The agreement immediately eliminated trade restrictions for many agricultural commodities. Protection for other commodities was scheduled to be phased out over time. The removal of these trade barriers was expected to encourage each country to specialize in the production of commodities in which they possess a comparative advantage.

Data presenting trade flows between Mexico and the USA is shown in Table 3.1. Since the implementation of NAFTA, Mexico has experienced an almost 50% increase in both fresh tomato and fresh vegetable exports to the USA. Clearly, Mexico's competitiveness in the US market has increased since NAFTA became effective. However, the USA not only continues to produce fresh tomatoes and fresh vegetables, but also even exports these products to Mexico. As presented earlier in this chapter, there are several reasons why complete specialization does not occur. A primary reason for this is increasing opportunity costs.

Table 3.1 Bilateral fresh tomato and fresh vegetable trade between Mexico and the USA, 1993–2000

	1993	1994	1995	1996	1997	1998	1999	2000
Bilateral fresh tomato trade between Mexico and the USA (Mt)								
US exports to Mexico	17,650	21,896	2,283	2,561	17,597	4,789	5,837	27,423
US imports from Mexico	400,494	376,032	593,063	685,678	660,609	734,053	615,064	589,954
Bilateral fresh vegetable trade between Mexico and the USA (Mt)								
US exports to Mexico	83,993	105,282	38,339	72,546	82,509	91,705	88,946	120,312
US imports from Mexico	1,383,003	1,392,944	1,817,526	2,104,991	2,097,545	2,270,783	2,138,737	2,074,979

Source: FATUS (2003), Economic Research Service, US Department of Agriculture, Washington, DC

■ 3.5 GAINS FROM TRADE

The total gain from trade can be divided into the following two components: (a) the gain from international exchange, or the consumption gain; and (b) the gain from specialization, or the production gain. In Figure 3.6, the curve KM is the economy's PPF. Before trade, equilibrium occurs at point P_0 (30 units of Y and 40 units of X) where the PPF is tangent to the highest attainable SIC (SIC_0). The common slope of the PPF and the SIC is the autarky price ratio $(P_x/P_y)^a$. After opening trade, with the terms of trade (P_x/P_y) lower than the autarky price ratio, the economy produces at P_e (40 units of Y and 30 units of X), where the PPF is tangent to an income line. This income line is also known as the CPF. Since the CPF is the highest income line the economy can reach, the economy obtains the maximum income at production point P_e.

Consumption equilibrium is obtained at C_e (28 units of Y and 53 units of X), where the CPF is tangent to the highest attainable social indifference curve. Social utility is improved because the economy can reach a higher level of the social indifference curves (SIC_1). A shift of the social indifference curve from SIC_0 to SIC_1 represents the total welfare gain from trade.

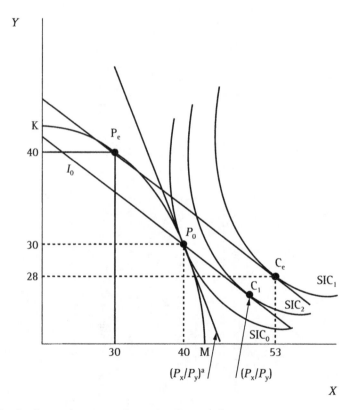

Figure 3.6 Production and consumption gains from trade.

To isolate the consumption gain, assume that the production point is frozen at P_0 after opening trade. The economy's income, when it produces at the autarky production point (P_0), is represented by income line I_0. The economy can consume at any point on the income line when it produces at P_0. In this case, the consumption equilibrium point is obtained at point C_1, where the income line I_0 is tangent to the highest attainable social indifference curve. This social indifference curve is the highest one that this economy can reach at the given production constraints. The economy reaches a higher social indifference curve (SIC_2) than the one associated with the autarky equilibrium (SIC_0). This implies that even though production is frozen at P_0, the economy still benefits from trade. The shift of the social indifference curve from SIC_0 to SIC_2 is the consumption or international exchange gain. When the economy moves production from P_0 to equilibrium production point P_e, the corresponding income line (CPF) is tangent to SIC_1 at consumption equilibrium point C_e. Thus, the movement of the SIC from SIC_2 to SIC_1 is the production or specialization gain.

■ 3.6 NEOCLASSICAL DEMONSTRATION OF COMPARATIVE ADVANTAGE

Assume that two countries, the United States and the United Kingdom, are endowed with fixed quantities of two inputs of production, labor (L) and capital (K), and produce two commodities, corn and textiles. The US pre-trade equilibrium is obtained at point P^a, where its PPF is tangent to a social indifference curve (Figure 3.7(a)). The common slope of the US PPF and social indifference curve is the US pre-trade equilibrium price ratio $(P_t/P_c)^{US}$. Similarly, the UK pre-trade equilibrium can be obtained at point P^b, where its PPF is tangent to a social indifference curve (Figure 3.7(b)). The common slope of the UK PPF and social indifference curve is its pre-trade equilibrium price ratio $(P_t/P_c)^{UK}$.

Assume that the pre-trade equilibrium price ratio in the USA is greater than that in the UK: $(P_t/P_c)^{US} > (P_t/P_c)^{UK}$. This implies that textiles are less expensive relative to corn in the UK and corn is less expensive relative to textiles in the USA before trade. The UK has a comparative advantage in the production of textiles and the USA has a comparative advantage in the production of corn. After opening trade, the USA starts to export corn to the UK and import textiles from the UK.

An increase in the supply of textiles through imports will reduce the price of textiles (P_t) in the USA and reduce the terms of trade from the pre-trade price ratio (P_t/P_c). Similarly, the UK starts to import corn from the USA. An increase in the supply of corn in the UK will reduce the price of corn (P_c) and the terms of trade will consequently start to rise from the pre-trade price ratio (P_t/P_c). Thus, the terms of trade should be determined somewhere between the US pre-trade price ratio and the UK pre-trade price ratio.

Let us look at the production and consumption equilibria in the two countries. For a given terms of trade, (P_t/P_c), lower than its pre-trade price ratio, $(P_t/P_c)^a$, the slope of the income line (the terms of trade) is flatter than that associated with the pre-trade price ratio. The production equilibrium point in the USA is obtained at P_e^a in

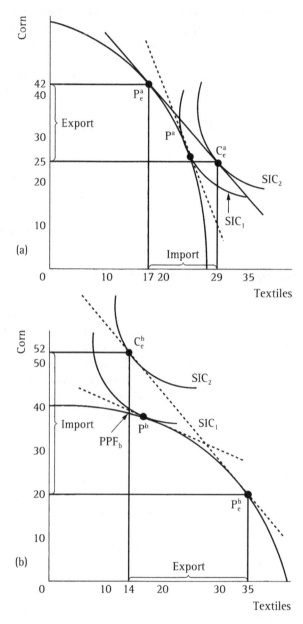

Figure 3.7 Production and consumption equilibrium points in (a) the USA and (b) the UK.

Figure 3.7(a), where its PPF is tangent to an income line, which is also known as the CPF. The consumption equilibrium point in the USA is obtained at C_e^a, where its CPF is tangent to a SIC. Consequently, the USA imports 12 yards of textiles and exports 17 tons of corn.

Similarly, for the UK, the slope of the income line (the terms of trade) is steeper than its pre-trade price ratio, $(P_t/P_c)^{UK}$. Consequently, production and consumption equilibrium points are identified at P_e^b and C_e^b, respectively, in Figure 3.7(b). The UK imports 32 tons of corn and exports 21 yards of textiles.

In this example, the USA exports corn and imports textiles. At the same time, the UK exports textiles and imports corn. Trade is mutually beneficial for both countries as long as the international price ratio lies somewhere between $(P_t/P_c)^{US}$ and $(P_t/P_c)^{UK}$.

■ 3.7 THE OFFER CURVE AND INTERNATIONAL EQUILIBRIUM

Assume that two countries, the USA and the UK, produce two commodities, textiles and corn, using two inputs, labor (L) and capital (K). Also assume that the USA has a comparative advantage in producing corn, while the UK has a comparative advantage in textiles. The US offer curve is defined as the amount of corn that the USA is willing to offer to the UK for different amounts of textiles. Similarly, the UK offer curve represents the amount of textiles that the UK is willing to offer to the USA for different amounts of corn. In other words, a country's offer curve represents optimal combinations of exports and imports associated with the corresponding terms of trade. Thus offer curves for these two countries can be derived from Figures 3.7(a) and 3.7(b).

In Figure 3.8(a), the y-axis represents US exports of corn, and the x-axis represents US imports of textiles. Lines passing through the origin represent terms of trade (TOT) of these two commodities. We can identify an optimal pair of exports of corn and imports of textiles associated with each TOT from Figure 3.7(a) and plot these levels of exports and imports in Figure 3.8(a). Point e_1 in Figure 3.8(a) represents an optimal combination of exports and imports associated with TOT_1. Similarly, points e_2 and e_3 represent optimal combinations of exports and imports associated with TOT_2 and TOT_3, respectively. The USA offers one unit of corn for increasing quantities of textiles as the TOT line rotates to the right – meaning that as the price of textiles becomes cheaper, the USA imports more textiles. We can derive the US offer curve by connecting the optimal combinations of exports and imports associated with the corresponding TOTs.

We can similarly derive the UK offer curve in Figure 3.8(b). This figure shows that the UK offers one unit of textiles for increasing quantities of corn as the TOT line rotates to the left – meaning that as the price of corn becomes cheaper, the UK imports more corn.

The offer curve passes through the origin, and the slope of an offer curve at the origin represents the pre-trade price ratio. The TOT_0 in Figures 3.8(a) and 3.8(b) represent the US and the UK pre-trade price ratios, respectively.

The international equilibrium is obtained at point E, where the US offer curve intersects the UK offer curve, as shown in Figure 3.9. TOT* represents the equilibrium terms of trade. At this terms of trade, the USA exports 80 tons of corn to the UK,

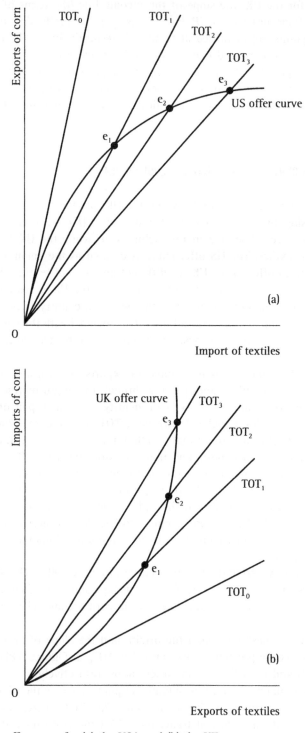

Figure 3.8 The offer curve for (a) the USA and (b) the UK.

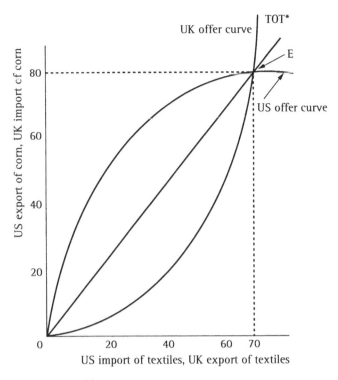

Figure 3.9 International equilibrium.

while the UK exports 70 yards of textiles to the USA. The US exports of corn are equal to the UK's imports of corn, and the UK's exports of textiles are equal to the US imports of textiles at the given TOT*.

SUMMARY

1   When two inputs (capital and labor) are introduced in the production of two commodities (X and Y), the production possibilities frontier becomes concave to the origin, mainly because these two inputs are not used in fixed proportion in the production of the two commodities. The concave PPF represents increasing opportunity costs, meaning that a country must give up more and more of one commodity to produce an additional unit of another commodity. The opportunity cost at a production point on the PPF can be measured by the slope of the PPF at that production point. The slope of the PPF is also known as the marginal rate of transformation (MRT).

2   In the absence of trade (autarky), a country's economy is in equilibrium where its PPF is tangent to its highest attainable social indifference curve. The common slope of the two curves at the tangency point gives the country's pre-trade equilibrium price ratio.

3   With trade, each country specializes in producing the commodity of its comparative advantage. The production equilibrium point in a country is obtained at a point on the PPF where

the PPF is tangent to its highest attainable income line, implying that the country can maximize its income when it produces at that production point. The income line tangent to the PPF is the consumption possibilities frontier (CPF), indicating that consumers in the country can consume at any point on the income line. The consumption equilibrium point in the country can be found at a point on the CPF where the CPF is tangent to its social indifference curve. At the consumption point, consumers in the country maximize their utility.

4   A country's offer curve is defined as the amount of commodity X that it is willing to offer to its trading partner for a different amount of commodity Y at a particular terms of trade. An offer curve passes through the origin, and the slope of an offer curve at the origin represents the pre-trade price ratio. The equilibrium conditions between the two countries can be obtained at the point at which the two countries' offer curves intersect.

5   Gains from trade can be divided into gains from exchange (consumption gains) and gains from specialization (production gains).

6   With a constant opportunity cost (linear PPF), a country completely specializes in producing a commodity for which the country has a comparative advantage over the other countries. However, specialization is partial with an increasing opportunity cost (concave PPF), because complete specialization becomes too expensive to accomplish.

KEY CONCEPTS

Autarky price ratio – A ratio of prices of two goods in a closed economy.

Consumption equilibrium – The quantities of two goods consumed to maximize consumers' utility.

Gains from exchange (consumption gains) – An increase in social welfare through trading goods between countries.

Gains from specialization (production gains) – An increase in social welfare through production specialization.

Marginal rate of transformation – The amount of one good that a country is willing to trade for another good.

Offer curve – A curve that represents the amount of a good that a nation is willing to trade to another nation for other goods at various prices.

Production equilibrium – The quantities of two goods produced to maximize producers' income.

QUESTIONS AND TASKS FOR REVIEW

1   Why is a production possibilities frontier concave to the origin?

2   When a production possibilities frontier is concave to the origin, why do opportunity costs of a commodity increase as more of the commodity is produced?

3   Why is there incomplete specialization in production with increasing opportunity costs?

4   What is meant by the production equilibrium point and how can it be derived?

5   What is meant by the consumption equilibrium point and how can it be derived?

6   What is meant by gains from exchange and those from specialization?

7   What is meant by the offer curve?

8   How can the international equilibrium be obtained?

SELECTED BIBLIOGRAPHY

Bhagwatti, J. N. 1964: The pure theory of international trade: a survey. *Economic Journal*, 74, 1–84.

Chacholiades, M. 1978: *International Trade Theory and Policy*. New York: McGraw-Hill.

Haberler, G. 1936: *The Theory of International Trade with its Applications to Commercial Policy*, trans. A. Stonier and F. Benham. London: Hodge. See Chs. 9 and 10.

Comparative Advantage and Factor Endowments: The Heckscher–Ohlin Theorem

■ **4.0 INTRODUCTION**

There are many reasons for trade between nations. Unlike the previous chapters, this chapter focuses on differences in resource endowments between nations as a reason for international trade. Countries have resource endowments in differing proportions. For example, Japan has very little agricultural land per person in comparison with the United States. China has a small amount of capital per person relative to either the USA or Japan. Given these differences in endowments, it is logical that countries will focus their production on those commodities that use their abundant resources most intensively. Since the USA has a larger relative endowment of land than Japan, the USA will likely export land-intensive commodities, such as beef, to Japan. Similarly, China will export labor-intensive goods to the USA and Japan, while the USA and Japan will export capital-intensive goods to China, since the USA and Japan have more capital per unit of labor than China.

A large amount of international trade theory involves the examination of endowments and their impact on international trade. The framework that shows the relationship between endowments and the pattern of trade is known as the endowment model. This chapter uses the endowment model to show the linkage between factor intensity, factor abundance, and the pattern of trade. The assumptions used in the Heckscher–Ohlin model are introduced and the Heckscher–Ohlin theorem is explained, along with an empirical test of the Heckscher–Ohlin theorem. An overview of the related factor-price equalization theorem and its implications for income distribution is also presented.

■ **4.1 ASSUMPTIONS OF THE THEOREM**

The Swedish economists Eli Heckscher and Bertil Ohlin formulated a theory explaining a nation's comparative advantage on the basis of factor endowments. The

Heckscher–Ohlin theorem, therefore, is also known as the factor-endowment theorem. This theorem states that comparative advantage is explained by differences in resource endowments such as labor and capital. For instance, China is an exporter of textile products which require intensive use of labor in their production process because China is heavily endowed with labor. At the same time, the USA exports computers and automobiles produced with intense use of capital because the USA has an abundance of capital.

The simple version of the theorem is based on the following simplifying assumptions:

1 Two countries use the same technology in the production of two commodities X and Y, meaning that the same production technologies are available in producing commodities X and Y in the two countries. If the prices of capital and labor were the same in both countries, producers in both countries would use the same relative amount of labor and capital in the production of the commodities.
2 Constant returns to scale are assumed in the production of the commodities. This implies that increasing the amount of labor and capital used in the production of any commodity increases the output of that commodity in the same proportion.
3 Factors of production are homogeneous between countries. This implies that factors used to produce commodities X and Y are the same in quality between the two countries.
4 Factors of production are perfectly mobile between industries within each country and are perfectly immobile between countries, implying that the factors of production cannot cross national boundaries.
5 There are no market distortions such as imperfect competition, labor unions, or taxes that would affect production or consumption decisions. In addition, transportation costs, tariffs, and other factors affecting the free flow of international trade are not allowed.
6 All resources are fully employed in producing commodities X and Y.
7 Social indifference curves are identical between countries, indicating that consumers in all countries have the same tastes and preferences.

With these assumptions, specialization of production continues until commodity prices are the same in both nations with trade. If we include transportation costs and tariffs, specialization would proceed only until relative commodity prices differed by no more than the costs of transportation and tariffs.

■ 4.2 FACTOR INTENSITY AND ABUNDANCE

The theorem is based on the concepts of factor intensity and factor abundance. It is important to clearly understand these concepts.

FACTOR INTENSITY

Commodity Y is defined as labor-intensive relative to commodity X if commodity Y uses relatively more units of labor (L) per unit of capital (K) than commodity X. Similarly, commodity Y is defined to be capital intensive relative to commodity X if commodity Y uses more capital per unit of labor than commodity X. For example, if two units of capital and one unit of labor are required to produce one unit of commodity X, the capital to labor ratio (K/L) is two. If at the same time one unit of capital and three units of labor are required to produce one unit of Y, the capital to labor ratio is 1/3. Since the capital to labor ratio in producing commodity X, $(K/L)_x$, is greater than that in producing commodity Y, $(K/L)_y$, commodity X is capital intensive relative to commodity Y, and commodity Y is labor intensive relative to commodity X.

Note that it is not the absolute amounts of capital and labor used in the production of X and Y that determine the factor intensity of the commodities, but the amount of capital per unit of labor (i.e., the capital to labor ratio, K/L). For example, assume that production of commodity X requires four units of capital and six units of labor and production of commodity Y requires two units of capital and two units of labor. Even though the production of commodity X uses more capital than that of commodity Y, commodity Y is capital intensive relative to commodity X, since the capital to labor ratio in producing commodity Y, $(K/L)_y$, is 1, which is greater than that in the production of commodity X, $(K/L)_x$, which is 2/3.

FACTOR ABUNDANCE

There are two ways to define factor abundance; one way is in terms of physical units of factors available in the two countries, and the other is in terms of relative factor prices. Country A is said to be labor abundant relative to country B if country A is endowed with more units of labor (L) per unit of capital (K) than country B. If the capital to labor ratio in country A, $(K/L)^a$, is greater than that in country B, $(K/L)^b$, then country A is capital abundant relative to country B. On the other hand, we can also evaluate factor abundance by factor prices. Assume that w and r represent the wage (price of labor) and the interest rate (the price of capital). The wage (w) is lower in a labor-abundant country relative to a capital-abundant country, while the interest rate (r) is lower in a capital-abundant country relative to a labor-abundant country. Country A is said to be labor abundant relative to country B if the wage to interest ratio in country A, $(w/r)^a$, is lower than that in country B, $(w/r)^b$.

If country A is labor abundant relative to country B and commodity X is labor intensive relative to commodity Y, then country A can produce relatively more of commodity X than can country B. At the same time, country B can produce relatively more units of commodity Y. This makes country A's production possibilities frontier (PPF) relatively flatter or skewed more toward the x-axis than that of country

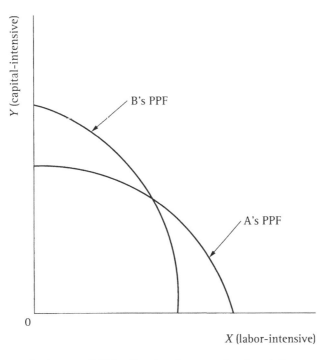

Figure 4.1 The production possibilities frontiers for countries A and B.

B, as shown in Figure 4.1. The properties of the PPF are the same as those discussed in the previous chapter. The PPF is concave to the origin, indicating that the opportunity cost of commodity X in terms of commodity Y increases as more units of commodity X are produced. The slope of the PPF at any point on the PPF represents opportunity costs.

■ 4.3 THE HECKSCHER–OHLIN THEOREM

The Heckscher–Ohlin theorem states that a capital-abundant country tends to specialize in the production of capital-intensive commodities and will export these commodities in exchange for labor-intensive commodities. A labor-abundant country tends to do the opposite.

Consider two countries, A and B, endowed with fixed quantities of two factors of production, labor (L), and capital (K), in producing commodities X and Y. Assume that country A is labor abundant relative to country B. Finally, assume that commodity X is labor intensive relative to commodity Y. According to the Heckscher–Ohlin theorem, country A has a comparative advantage in producing commodity X, and country B has a comparative advantage in producing commodity Y.

The Heckscher–Ohlin theorem indicates that differences in relative factor endowments between countries are the determinants of comparative advantage. For this reason, the Heckscher–Ohlin (H–O) model is referred to as the factor-proportions or factor-endowment model. For example, we can explain trade patterns between the USA and China using the H–O theorem. China is a labor-abundant country compared to the USA, and the USA is a capital-abundant country compared to China. Thus, China has a comparative advantage over the USA in producing labor-intensive commodities, and the USA has a comparative advantage over China in producing capital-intensive commodities. This implies that China exports labor-intensive commodities to the USA, and the USA exports capital-intensive commodities to China.

The H–O theorem is illustrated in Figure 4.2. The figure shows the PPF of countries A and B. Country A's production frontier is skewed toward the x-axis because

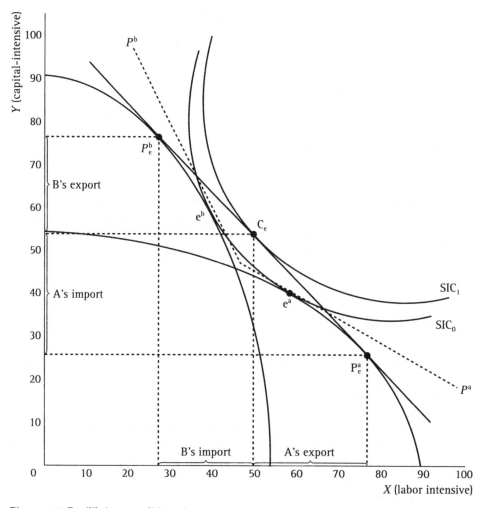

Figure 4.2 Equilibrium conditions in open economies.

the country is labor abundant and commodity X is labor intensive. On the other hand, country B's production frontier is skewed toward the y-axis because the country is capital abundant and commodity Y is capital intensive. Since the countries have identical tastes, they face the same set of social indifference curves (SIC_0). SIC_0 is the highest curve that countries A and B can attain in this example. Points e^a and e^b are the equilibrium points of production and consumption in the absence of trade in countries A and B, respectively. Before opening trade, country A produces and consumes 58 units of commodity X and 39 units of commodity Y, while country B produces and consumes 40 units of commodity X and 58 units of commodity Y. The slope of the PPF at points e^a and e^b defines equilibrium pre-trade price ratios in countries A (P^a) and B (P^b), which are equal to $(P_x/P_y)^a$ and $(P_x/P_y)^b$, respectively. Since the price ratio in country A (P^a) is less than in country B (P^b) in Figure 4.2, country A has a comparative advantage in producing commodity X, while country B has a comparative advantage in producing commodity Y.

After allowing trade, country A specializes in the production of commodity X and country B specializes in the production of commodity Y. Equilibrium production points, P_e^a and P_e^b, are found where the production possibility frontiers of countries A and B are tangent to the common price ratio line P^*. Country A produces 77 units of commodity X and 26 units of commodity Y, while country B produces 27 units of commodity X and 77 units of commodity Y.

Equilibrium consumption is obtained at the point C_e, where the price ratio line is tangent to a social indifference curve (SIC_1). Both countries consume 50 units of commodity X and 54 units of commodity Y. Country A produces at P_e^a on its PPF and consumes at C_e, indicating this country exports 27 (77 – 50) units of commodity X and imports 28 (54 – 26) units of commodity Y. Country B produces at P_e^b on its PPF and consumes at C_e, indicating that the country exports 23 (77 – 54) units of commodity Y and imports 23 (50 – 27) units of commodity X. Country A exports commodity X in exchange for commodity Y, and country B exports commodity Y in exchange for commodity X.

■ 4.4 EQUALIZATION OF FACTOR PRICES

The factor-price equalization theorem is based on the H–O theorem. Since the theorem was proven by Samuelson, it is sometimes referred to as the Heckscher–Ohlin–Samuelson (H–O–S) theorem.

The H–O–S theorem says that international trade will bring about equalization in relative and absolute returns to homogeneous factors across nations. This theorem implies that international trade is a substitute for the international movement of factors. For instance, international trade will cause the wages of homogeneous labor (i.e., labor with the same level of training, skills, and productivity) to be the same in all trading countries (countries A and B). Similarly, interest rates of homogeneous capital (i.e., capital of the same productivity and risks) will be the same among trading countries through international trade.

In the previous section, we assumed that the price ratio in country A (P^a) is lower than that in country B (P^b), implying that the price of commodity X is lower in country A than in country B. Since commodity X is labor intensive relative to commodity Y, a labor-abundant country, such as country A, specializes in the production of commodity X and reduces its production of commodity Y, as shown in Figure 4.2. The production specialization of commodity X in country A increases demand for labor, causing wages to rise, and the demand for capital decreases. Since a capital-abundant country, such as country B, specializes in the production of commodity Y, a capital-intensive commodity, the exact opposite occurs, leading to an increase in the interest rate and a fall in wages. Thus, international trade reduces the pre-trade differences in wages and interest rates between the two countries and will eventually equalize the prices of homogeneous factors. As long as relative commodity prices differ, trade between the two countries will continue. International trade expands until relative commodity prices are equalized. This implies that relative factor prices also become equal between the two countries. International trade also equalizes the absolute returns to homogeneous factors, under the assumptions discussed in the previous section.

Equalization of factor prices between two countries through international trade results in redistribution of income in the countries. Opening trade will raise the real wage and reduce the real returns to capital in the labor-abundant country, because the country specializes in producing labor-intensive goods (commodity X) and demands more labor and less capital. In the capital-abundant country, opening trade will raise the real returns to capital and reduce the real wage, because this country specializes in the production of capital-intensive goods (commodity Y) and demands more capital and less labor. This is known as the Stolper–Samuelson theorem. In general, changes in factor prices as a result of trade raise income levels in the two countries and change the distribution of income in the countries. In a labor-abundant country, labor income rises as a result of increases in wages, since the country specializes in the production of a labor-intensive commodity. On the other hand, capital income rises in a capital-abundant country since production specialization of a capital-intensive commodity increases demand for capital, resulting in an increase in the returns to capital.

■ 4.5 THE LEONTIEF PARADOX

The first empirical test of the H–O model was conducted by Wassily Leontief in 1951, using US data for the year 1947. Since the USA was the most capital-abundant country in the world, Leontief expected to find that the country exported capital-intensive commodities and imported labor-intensive commodities. In his study, Leontief estimated the K/L of US import substitutes rather than imports, mainly because foreign production data for actual US imports were not available. Import substitutes are commodities that the USA produces at home but also imports from abroad. Leontief argued that even though US import substitutes are more capital intensive than actual

BOX 4.1 THE RELATIONSHIP BETWEEN TRADE AND WAGES IN SOUTH KOREA

In 1970, South Korea was a labor-abundant country compared to the USA. Over 60% of its population were unskilled labor. Per capita income was about $267.00 per year in 1995 US dollars. As a result, the average hourly wage in South Korea was less than 2–3% of that in the USA. South Korea had a comparative advantage over the USA in producing labor-intensive goods, such as clothing and footwear. As a result, South Korea specialized in producing labor-intensive goods and exported them to the USA. South Korea exported $336 million to the USA in 1970; over 50% of its exports were labor-intensive goods. Increased exports of the labor-intensive goods increased demand for labor in South Korea, resulting in increases in the wages of workers in South Korea's textiles and apparel industries. Hourly wages increased about 30 times for the 1970–88 period (see Table 4.1).

Increased wages made the South Korean textiles and apparel industry less competitive in the US market, mainly because wages in other countries, such as China and India, were lower than in South Korea during the period. Ratios of South Korean exports of labor-intensive goods to its total exports to the USA fell from 60% in 1970 to 11.7% in 1993, and further decreased to 1.5% in 1998. On the other hand, wages in South Korea increased more than 100 times for the same period. This indicates that South Korea's textiles and apparel industry was losing its competitiveness to China and India in the US market. Average wages in South Korea are now only 10–20% lower than the US wage rate. This example demonstrates increases in wages in South Korea through its exports of labor-intensive goods to the USA.

Table 4.1 Bilateral trade in labor- and technology-intensive goods between the USA and South Korea and changes in income and wages in South Korea

	1970	1988	1993	1998
Per capita income (US dollars)	267	4,595	7,771	7,949
Wages (index 1995 = 100)	1.3	34.9	78.8	114.3
South Korean exports (million US dollars)				
Labor intensive	180	2,500	992	354
Technology/capital intensive	156	18,904	17,146	22,451
Total	336	21,404	18,138	22,805

Source: Korean Trade Information Service (KOTIS) database

imports, they should be less capital intensive than US exports if the H–O theorem held true.

Leontief found that US import substitutes were about 30% more capital intensive than US exports. This implies that the US exported labor-intensive commodities and imported capital-intensive commodities. This was the opposite of what the H–O model predicted and became known as the Leontief paradox.

In his study, Leontief tried to rationalize his results by including the concept of labor productivity. He argued that US labor was three times more productive than foreign labor, and that the US was a labor-abundant country if labor productivity was included when calculating the US labor force. He also argued that US consumers were biased so strongly in favor of capital-intensive commodities as to result in high relative prices for these commodities in the USA, resulting in increased imports of these capital-intensive goods. Another possible explanation presented by Leontief is that US imports are produced in foreign countries as labor intensive and similar goods are produced in the USA as capital intensive if the inputs are highly substitutable in producing the goods. The USA, a capital-abundant country, uses capital-intensive technology since wages are higher, while labor-abundant countries use labor-intensive technology since wages are lower. This is known as factor intensity reversal, which will be discussed more in the following section. However, these explanations of the Leontief paradox are not widely accepted.

More rational explanations of the Leontief paradox are that US tariff policy, which is biased against labor-intensive imports to protect US producers of similar products, results in increases in the production of import substitutes by using its abundant resource, capital. If there were no tariffs and trade restrictions on labor-intensive goods, the capital to labor ratio of US import substitutes would be lower. Another explanation of the Leontief paradox is that Leontief used a two-factor model (labor and capital), leaving out natural resources such as soil, mineral deposits, forests, and so on. Natural resources are important determinants of trade flows. Some US imports are natural resource intensive, and production of such products requires a larger amount of capital per unit of labor than imports that are not natural resource intensive. This explains the Leontief paradox.

In addition, Leontief used only physical capital and completely ignored human capital. Since US labor embodies more human capital than foreign labor, adding human capital to physical capital would make US exports more capital intensive relative to import substitutes. The USA is abundant in skilled labor and produces skilled labor intensive goods, exporting them. The recognition of human skill as human capital offers a partial explanation of the Leontief paradox.

■ 4.6 FACTOR INTENSITY REVERSAL

When two inputs, labor and capital, are highly substitutable in producing commodity X, this commodity could be labor intensive in the labor-abundant country A and capital intensive in the capital-abundant country B. This is a case of factor intensity reversal. Since wages are low in country A, this country will produce commodity X using labor-intensive techniques. On the other hand, country B will produce commodity X using capital-intensive techniques, since the wage is relatively high in the country. As a result, commodity X is a labor-intensive commodity in country A and a capital-intensive commodity in country B. However, if the substitution between labor and capital in the production of labor-intensive commodity Y is low, the two countries produce commodity Y using similar technologies.

When factor intensity reversals are present, neither the Heckscher–Ohlin theorem nor the factor-price equalization theorem holds. The H–O model fails because labor-abundant country A produces commodity X with labor-intensive techniques and exports commodity X, while the capital-abundant country B also produces commodity X with capital-intensive techniques and exports the commodity. Since the two nations produce and export the same homogeneous commodity, no trade equilibrium can be established between the two countries, meaning that the H–O model no longer predicts the pattern of trade. Since there is no trade between these two countries, factor prices are not equalized between the two countries.

SUMMARY

1   Commodity X is defined as labor intensive relative to commodity Y if commodity X uses more labor per unit of capital than does commodity Y. Similarly, commodity Y is defined as capital intensive relative to commodity X if commodity Y uses more capital per unit of labor than commodity X.

2   Country A is labor abundant relative to country B if country A is endowed with more labor per unit of capital than country B. Similarly, country A is capital abundant relative to country B if country A is endowed with more capital per unit of labor relative to country B.

3   The Heckscher–Ohlin theorem suggests that differences in nations' factor endowments are the basis for trade among nations. The theorem states that a capital-abundant nation has a comparative advantage in producing a capital-intensive commodity and a labor-abundant country has a comparative advantage in producing a labor-intensive commodity.

4   Contrary to the prediction of the Heckscher–Ohlin model, Leontief demonstrated in his empirical analysis that the USA, a capital-abundant country, exports labor-intensive goods and imports capital-intensive goods. His findings became known as the Leontief paradox.

5   The factor-price equalization theorem suggests that differences in prices of homogeneous factors between the two countries can be eliminated with trade. This implies that international trade can be a substitute for international mobility of factors, and that it equalizes absolute and relative prices of factors.

6   International trade also causes redistribution of income in a nation through the equalization of factor prices. Opening trade will raise the real wages, and reduces the real returns to capital in a labor-abundant country and vice versa in a capital-abundant country.

7   Factor intensity reversals occur when two inputs (e.g., capital and labor) are highly substitutable in producing a commodity. In this case, the commodity is labor intensive in a labor-abundant country and capital intensive in a capital-abundant country. Since the two countries produce and export the same commodity, the Heckscher–Ohlin theorem is no longer applicable to predict trade patterns between the two countries.

KEY CONCEPTS

Capital abundance – The quantity of capital available in a nation, relative to labor.

Capital intensive – The quantity of capital used to produce a good relative to the amount of labor.

Factor intensity reversal – A proposition indicating that a good can be capital intensive in a capital-abundant country and also labor intensive in a labor-abundant country if capital and labor are highly substitutable in producing the commodity.

Factor-price equalization theorem/Heckscher–Ohlin–Samuelson theorem – A proposition that indicates equalization of input prices between two countries through international trade.

Heckscher–Ohlin theorem/factor-endowment theorem – A proposition that indicates trade flows based on resource endowments between two countries.

Labor abundance – The quantity of labor available in a nation, relative to capital.

Labor intensity – The quantity of labor used to produce a good, relative to the amount of capital.

Leontief paradox – Empirical evidence that demonstrates trade flows that are inconsistent with the Heckscher–Ohlin theorem.

Stolper–Samuelson theorem – A proposition that indicates an increase in return of an input that is relatively abundant in a country and a decrease in return of other inputs through international trade.

QUESTIONS AND TASKS FOR REVIEW

1 State the assumptions of the Heckscher–Ohlin theorem. What is the meaning of these assumptions?
2 Explain the meaning of a labor-abundant country and a capital-abundant country.
3 Explain the meaning of a labor-intensive commodity, a capital-intensive commodity, and the capital to labor ratio.
4 Explain the trade pattern between the USA and China on the basis of the Heckscher–Ohlin theorem. It is generally assumed that the USA is capital abundant relative to China, while China is labor abundant relative to the USA.
5 What is meant by the Leontief paradox? What are some possible explanations of the paradox?
6 Explain the factor-price equalization theorem.
7 Explain the Stolper–Samuelson theorem.
8 What is meant by factor intensity reversal?

SELECTED BIBLIOGRAPHY

Baldwin, R. E. 1971: Determinants of the commodity structure of U.S. trade. *American Economic Review*, LXI(1), 126–46.

Heckscher, E. F. 1919: The effect of foreign trade on the distribution of income. *Ekonomisk Tidskrift*, XXI, 497–512. Reprinted in Ellis, H. S. and Metzler, L. A. 1950: *Readings in the Theory of International Trade*. Philadelphia: Blakiston, 272–300.

Keesing, D. B. 1966: Labor skill and comparative advantage. *American Economic Review*, LVI(2), 249–58.

Kravis, I. B. 1956: Wages and foreign trade. *The Review of Economics and Statistics*, XXXVIII(1), 14–30.

Leamer, E. E. 1980: The Leontief paradox reconsidered. *Journal of Political Economy*, 88(3), 495–503.

Leontief, W. 1951: Domestic production and foreign trade: the American capital position re-examined. *Economia Internazionale*, VII(1), 3–32. Reprinted in Caves, R. E. and Johnson, H. G. 1968: *Readings in International Economics*. Homewood, IL: Richard D. Irwin, 503–27.

—— 1956: Factor proportions and the structure of American trade: further theoretical and empirical analysis. *The Review of Economics and Statistics*, XXXVIII(4), 386–407.

Ohlin, B. 1967: *Interregional and International Trade*, revised edn. Cambridge, MA: Harvard University Press.

Samuelson, P. A. 1948: International trade and the equalization of factor prices. *Economic Journal*, LVIII(230), 165–84.

—— 1949: International factor-price equalization once again. *Economic Journal*, LIX, 181–97. Reprinted in Bhagwati, J. N. 1981: *International Trade: Selected Reading*. Cambridge, MA: The MIT Press, 3–16.

Stern, R. M. and Maskus, K. E. 1981: Determinants of the structure of U.S. foreign trade. *Journal of International Economics*, 11(2), 207–24.

CHAPTER 5

Imperfect Competition and Economies of Scale in Trade

■ 5.0 INTRODUCTION

Agriculture has traditionally been characterized as a purely competitive market. A large number of farmers produce and sell a commodity. Individual producers have no influence on the market. In other words, no individual farmer possesses market power. Given this industry structure, the traditional trade theories discussed in the previous chapters assumed pure competition and constant returns to scale.

As agriculture has developed, structural changes have occurred in the industry that make many of the previous assumptions questionable. Agribusiness now plays a much larger role in the industry. As a result, assumptions regarding pure competition may be too restrictive. In addition, the assumption of constant returns to scale is not realistic.

There has been a recent move to relax several of these assumptions, allowing for imperfect competition, economies of scale, and other characteristics that are more consistent with the real world. Economists such as Paul Krugman have pioneered the field known as *new trade theory*. Neoclassical trade theory considers differences such as resource endowments, technology, and preferences as the reason for countries to trade. These differences determine the comparative advantage between countries. New trade theory relaxes the assumptions of constant returns to scale and perfect competition, allowing for the analysis of situations involving economies of scale and imperfect competition.

■ 5.1 ECONOMIES OF SCALE AND INTERNATIONAL TRADE

Simplifying assumptions are often made to help analyze economic behavior. The models used in previous chapters have assumed constant returns to scale. This means that as inputs are changed by a certain percentage, outputs change by the same percentage. Depending on their size and level of production, firms may exhibit constant, increasing, or decreasing returns to scale.

ECONOMIES OF SCALE AND SPECIALIZATION

This can be seen using an example from production agriculture. At low levels of production, the yield of corn may be especially responsive to the amount of nitrogen used. Initially, increasing inputs by 10% may result in a greater than 10% increase in yield. This is an example of increasing returns to scale. Increasing inputs by another 10% may result in a 10% increase in yield. This is an example of constant returns to scale. Increasing fertilizer by another 10% may result in an increase in yield of less than 10%, or even decrease the yield. This is an example of decreasing returns to scale. The existence of economies of scale means that a proportional percentage increase in inputs will result in a larger percentage increase in outputs.

It is typically the case that efficiency increases with the size of the firm or industry. Consider the case of the agricultural machinery industry. Suppose that the United States and Canada have identical technologies for producing agricultural machinery. Each country can process 100 units of agricultural machinery with 100 units of labor. However, if either country doubles its input of labor to 200 units, it could produce 225 units of agricultural machinery given its input of capital and other resources.

Each country increases its productivity by concentrating on the production of a particular product. To maximize profits, countries will focus on the production of certain products. In this case, either the USA or Canada could specialize in the production of agricultural machinery, doubling its input in the industry and more than doubling the output of agricultural machinery. As each country specializes in a particular industry, resources are drawn away from other industries. This increases output, but without trade it may reduce the variety of products available for consumption. Trade allows countries to focus on the production of certain products and still maintain a variety of products for consumption.

Suppose that Canada now specializes in agricultural machinery, while the USA specializes in the production of Global Positioning System (GPS) equipment. Trade allows consumers in both countries to consume both goods despite the reduced production of certain goods in each country. Canada would export agricultural machinery in exchange for GPS equipment, while the USA would export GPS equipment in exchange for agricultural machinery. Total production increases and the combined consumption of the two countries increases.

Even though countries may have identical production technologies and resources, the existence of increasing returns to scale will encourage trade. By focusing production in strategic industries, countries can increase their efficiency more than if they tried to produce all commodities. Trade allows countries to increase their efficiency through specialization without losing the ability to consume a wide variety of products.

INTERNAL AND EXTERNAL ECONOMIES OF SCALE

While the concept of economies of scale has been discussed, it is important to recognize two ways in which economies of scale can be achieved. In determining how
international trade is impacted by economies of scale, it is essential to know what
type of economies of scale are presented in the industry. Efficiencies can be achieved
on the basis of either the size of the firm or the size of the industry. This introduces
us to the distinction between *internal* and *external* economies of scale. *Internal economies
of scale* are based on factors in the immediate control of the firm, while *external
economies of scale* are based on factors that are outside the control of the firm. In
general, internal economies depend on the size of the firm but not on the size of the
industry, while external economies depend on the size of the industry but not on the
size of the firm.

Consider the case in which an industry consists of 10 identical firms. Suppose that
the industry doubles in size and there are now 20 identical firms. This industry exhibits
external economies of scale if this increase in the size of the industry, but not in the
size of the firms, increases production efficiency. On the other hand, suppose that
the size of the industry remains the same, but there are now five identical firms in
the industry. Given that the size of the firms has doubled, this industry exhibits internal economies of scale if this increase in the size of the firms increases production
efficiency.

■ **5.2 IMPERFECT COMPETITION AND INTERNATIONAL TRADE UNDER INTERNAL ECONOMIES OF SCALE**

Production agriculture has historically been characterized by many small producers,
with no individual firm being large enough to influence the market through its actions.
This is the classic example of perfect competition. Firms in a perfectly competitive
market are price takers. A firm can sell as much as it wants, but it cannot influence
price. This occurs because each individual accounts for only a small share of total
production. Given this structure of the industry, the pure competition model has been
used to analyze agriculture and agricultural trade.

Through the course of development, the nature of agriculture has changed. In many
developed countries, there has been an outflow of labor from the agricultural sector.
As a result, there are fewer farmers. This creates the potential for imperfect competition, as the actions of individual producers could influence price. Over time, there
has been an increased amount of value added to food products at the post-farmgate
level. Given this, the role of agribusiness within the agricultural sector has changed.
This changing structure of the agricultural industry indicates that agricultural trade
may be better described through the use of imperfect competition. Imperfect competition is the situation in which firms are able to influence the market through the
amount that they sell or through price. Imperfectly competitive industries are composed
of either a small number of firms or firms that are able to create a differentiated

product. These firms are often referred to as price setters. There are many more examples of imperfect competition at the pre- and post-farmgate levels than in production agriculture.

MONOPOLISTIC COMPETITION AND COMPETITIVENESS

One example of imperfect competition is the monopoly. A monopolist has no rivals and faces no competition. The firm sets the price in order to maximize profits. However, the profit accruing to the monopolist encourages other firms in search of profit to enter the market. Because it is usually difficult to restrict entry, industries with economies of scale typically consist of several large firms that can influence prices but do not have complete control of the market. This type of industry, with several large firms, is the oligopoly market structure.

The existence of an oligopoly market structure creates several difficulties for examining the industry. Firms within an oligopoly can adjust their selling price in order to influence sales. But because of the relationship between firms in an oligopoly, their actions will affect the market price. In turn, this will influence the actions of other firms in the market. This strategic interdependence makes it difficult to develop a simple model that describes this behavior.

One type of oligopoly that has been used to avoid many of the problems associated with strategic interdependence is monopolistic competition. With monopolistic competition, firms can differentiate their product from that of their rivals. This insulates the firm from the actions of other firms, as consumers do not immediately switch loyalties because of one firm's behavior. The firm behaves as if its actions have no effect on other firms. This implies that firms take the prices of their competitors as given. Within the monopolistic competition framework, each firm behaves as though it has market power even though it is operating within a competitive environment.

Consider this behavior resulting from monopolistic competition. With monopolistic competition, a firm sells more as total demand for the industry's products increases and the price charged by rivals increases. Likewise, a firm sells less as the number of firms in the industry increases and its own price increases. The behavior of these firms determines the number of firms in the industry and their product price. Once these are determined, we can analyze how they are influenced by trade.

To accomplish this, consider the relationship between the number of firms and the average cost of production. As the number of firms increases, firm size will decrease. The existence of increasing returns implies that average cost of production will increase. This correlation between number of firms and cost of production is represented by the upward-sloping curve C in Figure 5.1.

The other relationship to consider involves the number of firms and product price. Competition increases with the increased number of firms. As the number of firms in the industry increases, prices are driven downward. This correlation between number of firms and sales price is represented by the downward-sloping curve P in Figure 5.1.

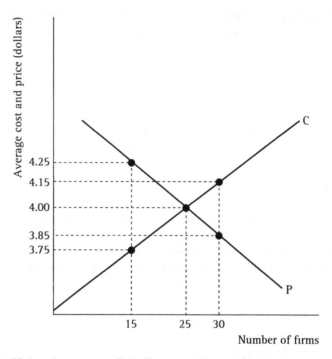

Figure 5.1 Equilibrium in a monopolistically competitive market.

Consider the firm's response, given alternative numbers of firms in the industry. In Figure 5.1, suppose that there are 15 firms in the industry. At this point, the sales price of $4.25 is greater than the average cost of $3.75. This creates an incentive for firms to enter the industry. Because of these excess profits, firms will continue to enter the industry until the sales price equals average cost. This occurs with 25 firms. Similarly, suppose that there are 30 firms in the industry. At this point, the average cost of $4.15 is greater than the sales price of $3.85. This creates an incentive for firms to exit the industry, given that average cost exceeds the sales price. Because of these losses, firms will continue to exit the industry until the sales price equals average cost. This occurs, once again, with 25 firms.

The intersection of the cost and price curves corresponds with 25 firms. With 25 firms in the industry, the average cost of $4.00 is equal to the sales price of $4.00. This indicates that the zero-profit number of firms for this industry is 25. More than 25 firms in the industry will result in losses, encouraging firms to leave the industry. Fewer than 25 firms in the industry will result in profits, encouraging firms to enter the industry. The long-term equilibrium number of firms in this industry is 25 and the equilibrium price is $4.00. This result can now be used to examine the impact of monopolistic competition on international trade.

MONOPOLISTIC COMPETITION AND INTERNATIONAL TRADE

Without trade, consumption is limited by domestic production. Likewise, production can only be sold within the domestic market. As countries are able to trade with one another, these limits on production and consumption are removed. International trade has the effect of increasing the size of markets. Given the framework developed in the previous section, we will examine the impact that international trade can have on an industry.

Take the case of two countries, where farmers in each country purchase 1,000 center-pivot irrigation systems and 500 units of farm machinery every year. Through specialization resulting from trade, the size of the market increases to 2,000 units of center-point irrigation systems and 1,000 units of farm machinery, creating opportunities for firms to produce products at a lower average cost. This concept can be analyzed using monopolistic competition as presented earlier. Figure 5.2 shows the effect of increased market size on price and number of firms. Consider the initial relationship between average cost and number of firms in the industry prior to trade. Given the relationship between price and number of firms, an initial equilibrium occurs at point a, with five firms producing center-pivot sprinklers at a cost of $10,000 per unit. As shown earlier, this indicates that the zero-profit number of firms in the industry is five.

Now suppose that the size of the market increases because of trade. Because of increased market size, firms are better able to take advantage of internal economies of scale. For any fixed number of firms in the industry, an increase in total sales will reduce average cost. This results in a downward shift in the average cost curve

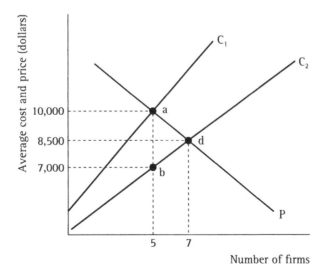

Figure 5.2 Effects of increased market size through trade.

to C_2. Holding the number of firms in the industry constant at five, excess profits would exist in the industry, as the price charged ($10,000) is greater than the average cost of production ($7,000). This creates the incentive for the entry of additional firms into the industry. As the number of firms increases, prices will go down and average cost will increase, until a new zero-profit equilibrium price of $8,500 is reached at point d, with seven firms in the industry. At the same time, farmers pay a lower price for sprinkler irrigation systems. Monopolistic competition results in the production of a variety of differentiated products. Given the increase in variety and decrease in price, it is clear that farmers gain from these developments.

IMPERFECT COMPETITION AND THE PATTERN OF TRADE

The previous discussion presented an economic condition that demonstrated how the existence of monopolistic competition can result in gains from trade even though there may not be any differences between countries with respect to endowments or technology. However, it said little about how the pattern of trade would develop. In other words, it could not determine whether all of the industry's firms would be located in the home country, in the foreign country, or split between the two. Consider how the pattern of trade is determined.

Suppose that China and the USA each possess capital and labor that can be used to produce two goods, machinery and textiles. The USA is capital abundant, while China is labor abundant. In addition, the machinery industry is capital intensive, while the textile industry is labor intensive.

Let us utilize factor proportions theory to review the flow of goods between the countries due to comparative advantage. Given that the USA is capital abundant, it will export capital-intensive machinery and import labor-intensive textiles. Likewise, since China is labor abundant, it will export labor-intensive textiles and import capital-intensive machinery. This flow of goods, or pattern of trade, is the basic exchange of goods between the two countries.

Suppose that the market structure of the textiles industry is monopolistic competition, in which each firm produces a differentiated product. As indicated earlier, because of internal economies of scale, neither country produces the complete range of textile products by itself. Because both countries are producing some textile products, they specialize by producing different textile products.

The characterization of the textiles industry as monopolistically competitive gives rise to the situation in which China is still an exporter of textiles and an importer of machinery, but also imports textiles as well. Given the desire for textile products produced in the USA, China imports some textile products from the USA.

The consideration of monopolistic competition in analyzing the pattern of trade is important, because it highlights a real-world phenomenon. Most countries import and export products in a manner characterized by intra-industry trade under monopolistic competition. When economies of scale are considered, the pattern of trade is influenced by both comparative advantage and economies of scale. The USA continues to export capital-intensive machinery and import labor-intensive textiles. This exchange of machinery for textiles is referred to as inter-industry trade. However, the USA also exports some labor-intensive textiles to China. This exchange within an industry – textiles for textiles – is an example of intra-industry trade.

There are several key points from this example. First, the existence of inter-industry trade is based on comparative advantage. The comparative advantage of countries is influenced by their endowment of resources and level of technology. These differences result in inter-industry trade. On the other hand, the existence of intra-industry trade is not necessarily the result of comparative advantage but may occur as a result of other factors, including economies of scale and imperfect competition.

BOX 5.1 THE US SOYBEAN EXPORT EMBARGO

Increases in the price of soybeans in the early 1970s prompted the USA to impose an embargo on soybean exports. This policy was designed to control record high domestic soybean prices that resulted from increased European consumption and overall world demand. The embargo was successful in lowering domestic prices from their high of $10 per bushel. It was eventually replaced by an export licensing system that covered soybeans and soybean-related products.

While these policies of the USA were successful in moderating an increase in the domestic price of soybeans, the world soybean prices increased dramatically in the weeks following the implementation of the embargo. In hindsight, achieving the short-term policy objective of moderating domestic prices is of small consequence compared to the long-term impact that the embargo may have had on the world soybean market.

The impact of these actions took the form of decreased confidence in the dependability of the USA as a soybean supplier. Perhaps the greatest damage occurred with respect to Japan, a major consumer of soybeans. As a result of the embargo, Japan identified alternative suppliers of soybeans and soybean products. The most significant development involved Japan's investment in and collaboration with Brazil as an alternative supplier. This caused Brazil's production infrastructure to be enhanced, resulting in an increased ability to compete in the world soybean market.

Brazil's soybean industry was in its infancy in the early 1970s, supplying just over 2% of the world export market, compared to a 93% market share for the USA (see Figure 5.3). Japanese investment, combined with expanding market opportunities, provided the catalyst necessary for the Brazilian soybean industry to develop. As a result, Brazil's share of the world soybean market has increased. In 2000, its market share was over 24%, while the US share had declined to 57%. Although the comparative advantage of Brazil had not changed, the US soybean embargo provided the impetus for the Brazilian soybean industry to grow into a competitive player in the world soybean market.

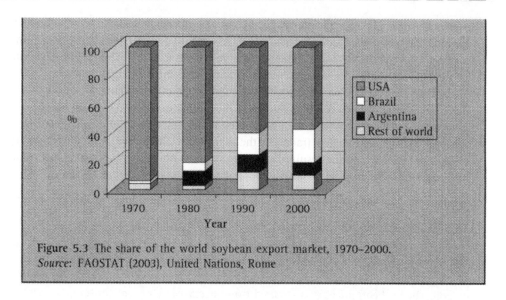

Figure 5.3 The share of the world soybean export market, 1970–2000.
Source: FAOSTAT (2003), United Nations, Rome

It is important to note that the pattern of trade between countries is influenced to a great extent by their relative similarity. In general, intra-industry trade is more important between countries that are relatively similar in resource endowments, while inter-industry trade is more important between countries that are relatively different in resource endowments. For example, the USA and Canada have many similarities with respect to endowments of resources and technology. As a result, a large amount of intra-industry trade takes place between these two countries. One example of intra-industry trade involves the agricultural machinery industry, where a significant amount of trade in agricultural machinery and machinery parts takes place between the two countries. The opposite would hold true for the USA and Mexico. Significant differences in resources, technology, and climate dictate that the pattern of trade be more specialized, leading to inter-industry trade. As a result, Mexico exports large quantities of tropical agricultural products to the USA in exchange for manufactured goods.

■ 5.3 EXTERNAL ECONOMIES AND INTERNATIONAL TRADE

To this point, the discussion has considered the impact of internal economies of scale in determining the pattern of trade. The case has been made that internal economies of scale encourage firms to increase in size and thereby lower their average cost. As firm size increases, so does the likelihood of imperfect competition.

EXTERNAL ECONOMIES AND EFFICIENCY

There is another categorization of scale economies that influences an industry's efficiency, in addition to internal economies of scale that are specific to the individual firm. External economies influence production efficiencies through factors at the level of the industry. This can be seen through the observation that certain industries have tended to develop in clusters. One example of this type of clustering in agriculture occurred in the late nineteenth and early twentieth centuries, as the US meat-packing industry was heavily concentrated in the Chicago area.

External economies may or may not be related to comparative advantage. It is clear that certain factors associated with comparative advantage can encourage the development of an industry in one location rather than another. But happenstance is often a key factor. Take, for example, the case of Hollywood, California. In choosing the location for his movie studio, Cecil B. DeMille had narrowed the choices down to Tucson, Arizona, and Hollywood, California. There were likely some key factors that influenced the final decision to locate in California. However, once that decision was made, the development of the movie industry in Hollywood during the twentieth century produced a tremendous infrastructure. The industry that is concentrated in this central location is much more efficient than any individual movie studio located in isolation.

There are several ways in which this clustering of firms can influence the efficiency of an industry. First, the close proximity of a large core group of firms encourages the emergence of ancillary firms that provide support to the industry. This development of local specialized suppliers has the effect of reducing cost within the industry. Second, a high concentration of firms in the same industry results in a pool of labor specifically trained for the needs of the industry. From the firm's perspective, an adequate supply of labor is available. This reduces the cost associated with finding additional labor as well as in training employees. For workers, the cost associated with moving from one job to another declines, from a perspective of relocation and down time. Third, interaction among employees from the same industry will increase the likelihood of knowledge spillovers. The sharing of knowledge and advances in technology will benefit the entire industry, both through current production gains and as they act as a catalyst for future advances and developments.

This overview of factors that give rise to external economies of scale suggests that firms in close proximity tend to gain production efficiency relative to their isolated counterparts. As the size of an industry within a nation increases, so does the opportunity and incentive for clustering. When an industry is characterized by external economies, the country with a large industry will likely possess a cost advantage over the country with a small industry. This leads to the concept of the "forward-falling" supply curve. This refers to the phenomenon that, as the size of the industry increases, the industry's average cost declines.

EXTERNAL ECONOMIES AND INTERNATIONAL COMPETITIVENESS

The forward-falling supply curve resulting from external economies of scale has interesting implications for international trade patterns. External economies of scale result in efficiency gains as the size of the industry increases. However, the evolution of an industry within a country is influenced by a variety of factors. We have seen that the productivity of an industry is affected by endowments and technologies, but other factors such as domestic policies and circumstance can be major determinants in the development and growth of an industry.

Consider the situation of an industry in which two countries exhibit external economies of scale. In both countries, the relationship between quantity produced and average cost is represented by a forward-falling supply curve. We will consider this situation using the case of the US and the German motion picture industries, depicted in Figure 5.4. For a variety of reasons, the US motion picture industry has a much more developed infrastructure than its German counterpart. Because of this, an infrastructure exists in the USA whereby a number of firms related to movie production have been established within close proximity. These include actors, talent agencies, production companies, and many more related firms that add to the overall strength of the industry. As firms of this type are established, the industry's efficiency increases. This serves to increase the size of the industry as it takes advantage of external economies. The US cost of producing motion pictures is represented by AC_{US} in Figure 5.4.

Germany also has the capability to produce motion pictures. In fact, given endowments related to its ability to efficiently produce German-language films, Germany is actually better suited to produce this product than the USA. This is depicted by Germany's cost of producing motion pictures, AC_{GER}, which lies below AC_{US}. This indicates that Germany displays a cost advantage over the USA when both are producing the same quantity.

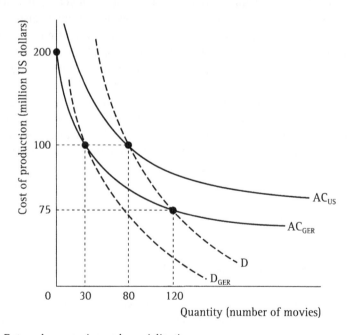

Figure 5.4 External economies and specialization.

If the industries were not characterized by external economies of scale, individual firms could begin operations in Germany to compete with the USA in the production of motion pictures. However, given that the size of the industry influences production efficiencies in this example, the initial position of the USA as the dominant country makes it difficult for Germany to enter the market.

Given world demand for motion pictures, represented by D, the USA as the dominant country produces 80 motion pictures per year, at a cost of $100 million per movie. If Germany established its capacity, it would be able to produce 30 movies per year at a cost of $100 million per movie. This would allow Germany to equal the price charged by the US industry, thus challenging the USA as the dominant world motion picture producer.

This prospect may not be so easy, however. The USA has been entrenched as the dominant producer for a number of years. Given that these industries exhibit external economies, the US production capacity may be an insurmountable obstacle for Germany to overcome. With external economies, gains in efficiency are not associated with individual firms, but with the growth of the industry as a whole.

From the initial point of no production, the German industry can produce movies at a cost of $200 million. This is well above the initial equilibrium cost of $100 million to produce movies in the USA. To compete with the USA, Germany must increase the size of its industry to the point at which it is capable of producing 30 movies per year. Germany's ability to enter the market depends on the US motion picture industry's level of establishment. Establishing the German industry to this level would be a difficult task.

This example creates an interesting situation. If Germany's domestic demand for movies is sufficiently large, it is possible that the country can simultaneously develop its industry and provide gains to consumers by restricting trade or by providing production subsidies. Referring once again to Figure 5.4, suppose that Germany's domestic demand is represented by D_{GER}. Suppose that imports of US movies are restricted by means of an import tariff or quota. Given sufficient time and incentive to expand, the German motion picture industry will be able to meet its domestic demand, producing 30 motion pictures per year at a cost equal to that of the US industry. Once Germany's industry has reached this stage, it can be allowed to develop without further protection from the government. This would ultimately allow Germany to expand its industry to the point at which it can produce 120 movies per year at a cost of $75 million per movie, replacing the USA as the dominant producer.

External economies of scale can also result in the phenomenon of dynamic increasing returns, often referred to as the learning curve. The previous example showed that the supply curve depends on the current level of production. The learning curve shows the relationship between cumulative production and production efficiency. An industry that has been in operation for a long period of time and takes advantage of external economies, such as the agricultural chemical industry in the USA and Europe, becomes entrenched as the number of firms becomes geographically concentrated. Over time, the expertise of the firms is enhanced through knowledge spillover among firms. This knowledge accumulates such that the longer the industry exists,

BOX 5.2 ECONOMIES OF SCALE AND IMPERFECT COMPETITION IN AGRICULTURE: THE CASE OF THE CANADIAN AND US BEEF INDUSTRIES

The Canadian and US beef industries are both characterized by large processing facilities that convert live cattle into boxed beef. To compete in this market, plants must be large enough to process thousands of cattle per day. As opposed to the late nineteenth and early twentieth centuries, when processing facilities were located near railway terminals, today's beef processors are located in areas with heavy concentrations of cattle feedlots.

The USA is clearly the larger of the two industries, producing nearly 10 times Canada's production, as shown in Table 5.1. However, Table 5.2 shows that the value of US beef and live animal imports from Canada is much more than the value of US beef and live animal exports to Canada. Why does such a large *producing country* import so much product from a producer one-tenth its size? The answer may lie in economies of scale and intra-industry trade.

Several major US beef processors, such as Excel and IBP, have boxed-beef facilities in Canada. There are many similarities between the USA and Canada with respect to beef processing, so it is not clear that either country has a comparative advantage. When economies of scale exist, countries can specialize in a particular area through engaging in international trade. This allows production efficiencies to be gained as countries take advantage of economies of scale.

Table 5.1 Beef and veal production in Canada and the USA, 2002 (in metric tons)

	Canada	USA
Production	1,290,000	12,438,000

Source: FAOSTAT (2003), United Nations, Rome

Table 5.2 The flow of the beef trade between Canada and the USA, 2002 (in thousands of US dollars)

	Beef and veal	Cattle and calves	Total
Imports from Canada to the USA	1,113,220	1,145,766	3,372,206
Exports from the USA to Canada	286,282	50,003	622,568

Source: FATUS (2003), Economic Research Service, US Department of Agriculture, Washington, DC

One potential scenario is as follows. Both countries could produce feeder calves, fed cattle, and boxed beef at the same level of productivity. If the USA shifts its emphasis to the production of feeder calves, it may gain production efficiencies. In addition, since it is assumed that these economies of scale for feeders and processors are internal as opposed to external, efficiency gains depend upon the size of a particular feedlot or processing plant. There might be a cluster of feedlots around a plant, but another processor does not gain efficiencies by locating adjacent to other plants.

Because the USA has gained efficiencies in the feeder cattle sector, intra-industry trade (trade within the beef industry) has increased. The USA becomes more efficient in producing feeder calves, using a part for feeding and processing and exporting the remainder to Canada. Because it is internal economies that exist in processing and feeding, the USA does not lose efficiencies by Canada focusing in these areas. Canada imports feeder calves from the USA to supplement its own production. These animals are fed and processed, with a significant amount of its boxed-beef production exported to the USA.

Other factors are at play in addition to economies of scale. Other reasons for shifting production facilities from one country to another include differences in import tariffs, corporate taxes, input costs, and the market power of input suppliers. Another factor is the market power of the processors. The two US firms mentioned earlier, Excel and IBP, account for a large portion of the US beef processing market as well as the Canadian market. As large firms dominate an industry, their behavior may result in trade flows such as those observed in the US–Canada beef industry.

As companies gain market power, they also gain the ability to restrict the entry of new firms into the industry. This is especially true when internal economies of scale are present. Significant amounts of resources are needed to obtain and operate a beef processing plant with the capacity necessary to compete with existing processors. As a dominant firm *moves* from controlling an industry in one country to controlling the industry in two countries, there will likely be a tendency for the firm to make decisions based on the combined market. Large firms with market power will likely assess the entire situation to determine optimal plant location. When economies of scale exist, *firm-level* decisions may result in an increase in intra-industry trade rather than a complete elimination of production in one country at the expense of complete specialization in the other.

the greater is its advantage. In general, this is quite similar to the concept of the forward-falling supply curve. However, there is an important distinction. While the forward-falling supply curve depends on the current and potential level of output, the learning curve is influenced by the total level of knowledge gained as the result of cumulative output.

■ 5.4 STRATEGIC TRADE POLICY, ECONOMIES OF SCALE, AND IMPERFECT COMPETITION

The existence of imperfect competition and economies of scale challenges several basic assumptions of neoclassical trade theory. Increasing returns to scale will encourage trade even though countries may have identical production technologies, resources, and tastes. In addition, imperfect competition can result in gains from trade even though endowments and technology are the same across countries. In both cases, strategic planning can allow countries to increase their efficiency through specialization.

Achieving specialization of this type can be difficult. As seen in the case of external economies of scale, it is hard to predict the emergence of industries, let alone direct it. Yet as policy-makers seek to maximize societal welfare, the ability to direct the growth of industries is critical. One important possibility is that countries can strengthen industries through protection, thus enhancing their welfare.

An example of this involves the development of strategic trade policy to enhance the competitive position of key firms or industries. An example of this involves Japan during the second half of the twentieth century. Often referred to as Japan Inc., Japan was able to use investment strategies combined with protectionist policies to develop industries deemed key to the economic growth and welfare of the country. The Japanese automobile and electronics industries are now world leaders. Strategic trade policy contributed to their development. Because these industries are characterized by imperfect competition and exhibit increasing returns to scale, the use of trade protection can result in increased market power and efficiency through specialization.

As countries recognize opportunities to achieve this type of specialization through protection, the pressure to insulate industries from the world market may increase. While numerous agreements call for freer trade throughout the world, these pressures create a dilemma for policy-makers, as they are torn between honoring trade treaties and enhancing strategic industries. An example of this occurred in 2003, when the USA imposed tariffs on steel imports. This action was in direct violation of WTO agreements and resulted in sanctions being imposed on goods produced in the USA. However, the USA implemented the tariff despite any knowledge it may have had concerning the tariff's legalities, indicating that the benefits of the tariff may have been greater than the sanctions and other costs.

While most economists recognize the gains that can be achieved through free trade, many also realize that there are conditions in which protection can be beneficial. The existence of imperfect competition and economies of scale are two such situations. As the world becomes increasingly global and the role of large multinational firms increases, so too will the pressure to provide protection to key industries. As this occurs, the incentive to participate in free trade agreements will be affected. Many countries will elect to protect their strategic industries, yet will also seek foreign market access for their domestic production. As the use of this type of protectionist policies increases, it may become more difficult to achieve progress in trade negotiations.

SUMMARY

1   Many industries exhibit increasing returns to scale. It is typically the case that efficiency increases with the size of the firm or industry. The existence of increasing returns to scale means that a percentage increase in inputs will result in a larger percentage increase in outputs. It is suggested that economies of scale are one reason for trade. By limiting production to a specific set of industries, a country can raise its level of efficiency to a point that is higher than if it tried to produce all commodities. In this manner, countries are able to increase their efficiency through specialization without losing the ability to consume a wide variety of products.

2   Differences in the way in which reductions in average production costs are achieved highlight an important distinction concerning economies of scale. Efficiency gains can be achieved through economies of scale in two ways. *External economies of scale* occur when the cost of production depends on the size of the industry. *Internal economies of scale* occur when the cost of production depends on the size of the firm.

3   The existence of an oligopoly market structure can create several difficulties for research and analysis. One of the main problems results from the level of interdependence between firms within the oligopoly market structure. The resulting strategic interdependence makes it difficult to develop a simple model that captures this behavior.

4   In industries where there are economies of scale, limitations imposed by the size of the market can constrain both the variety of goods produced and the size of firms. Increasing the size of the market through trade removes these limitations, allowing each country to specialize in a narrower range of products in order to gain efficiencies through internal economies of scale. At the same time, since other countries are able to gain efficiencies through internal economies of scale, the reduction in variety caused by specialization is compensated for through trade. Through purchasing goods that it does not produce, each country can increase the variety of goods available for consumption. Gains from trade can occur even though countries may not differ in their endowments of resources or technology.

5   The existence of inter-industry trade is a result of comparative advantage. The comparative advantage of countries is influenced by their endowment of resources and level of technology. These differences result in inter-industry trade. On the other hand, the existence of intra-industry trade is not a result of comparative advantage but occurs as a result of other factors, one of which is the existence of internal economies of scale.

6   As opposed to internal economies, external economies influence the average cost of production through factors at the level of the industry rather than the firm. This can be seen through the observation that certain industries have tended to develop in clusters. Examples of this include the concentration of the US film industry in Hollywood, the investment banking industry on Wall Street, and the semiconductor industry in Silicon Valley.

7   The forward-falling supply curve resulting from external economies of scale has interesting implications for international trade patterns. However, the evolution of an industry within a country is influenced by a variety of factors. We have seen that the productivity of an industry is affected by endowments and technologies, but other factors such as domestic policies and circumstance can be major determinants to the development and growth of an industry. External economies of scale can also result in the phenomenon of dynamic increasing returns, also referred to as the learning curve. While the forward-falling supply curve depends on the current level of output, the learning curve is influenced by the total level of knowledge gained as the result of cumulative output.

KEY CONCEPTS

External economies of scale – The impact of factors that are outside the control of the firm, such as industry size, on a firm's production efficiency.

Forward-falling supply curve – The relationship between the current level of production and cost in which increased current production results in decreased cost.

Imperfect competition – The situation in which firms are able to influence the market through the amount that they sell or through price.

Increasing returns to scale – The situation in which a proportional percentage increase in inputs will result in a larger percentage increase in output.

Inter-industry trade – The exchange of goods between two countries on the basis of the principle of comparative advantage.

Internal economies of scale – The impact of factors in the immediate control of the firm, such as firm size or level of production, on the firm's production efficiency.

Intra-industry trade – The exchange between two countries of similar goods within the same industry.

Market size – The quantity of product that could potentially be bought by consumers who are able to purchase a firm's product.

Monopolistic competition – A type of oligopoly in which firms can differentiate their product from that of their rivals.

Oligopoly – The type of market structure in which there are several large firms that can influence prices but do not have complete control of the market.

Strategic trade policy – The development of public or private programs designed to enhance the competitive position of key firms or industries.

The learning curve – The relationship between the cumulative level of production and cost in which increased cumulative production results in decreased cost.

QUESTIONS AND TASKS FOR REVIEW

1 How could the existence of increasing returns to scale impact decisions by the government to promote growth in one industry versus another? How does this affect the belief that trade protection results in welfare losses?

2 Provide an example of external economies of scale and an example of internal economies of scale. How might the behavior of a profit-maximizing firm with external economies of scale be different from that of one that possesses internal economies of scale?

3 How does the existence of an oligopolistic market structure and the resulting strategic interdependence between firms make the modeling of economic behavior more difficult?

4 Explain how trade allows a country to specialize in the production of fewer goods, yet gives consumers access to a greater variety of goods. How does this specialization in the production of fewer goods allow the countries to gain efficiencies through economies of scale?

5 What is the difference between inter-industry and intra-industry trade? Explain.

6 How can the existence of a forward-falling supply curve or a learning curve give governments incentive to provide protection to specific firms or industries? Is this a valid argument for protection?

SELECTED BIBLIOGRAPHY

Arrow, K. 1962: The economic implications of learning by doing. *Review of Economic Studies*, 29, 153–73.

Bain, J. 1954: Economies of scale, concentration and entry. *American Economic Review*, 44, 15–39.

Baumol, W. 1962: On the theory of expansion of the firm. *American Economic Review*, 52, 1078–87.

Chamberlin, E. 1933: *The Theory of Monopolistic Competition*. Cambridge, MA: Harvard University Press.

Dixit, A. and Stiglitz, J. 1977: Monopolistic competition and optimum product diversity. *American Economic Review*, 67, 297–308.

Krugman, P. 1995: Increasing returns, imperfect competition and the positive theory of international trade. In G. M. Grossman and K. Rogoff (eds.), *Handbook of International Economics*, vol. 3. Amsterdam: North-Holland.

—— and Obstfeld, M. 2000: *International Economics: Theory and Policy*, 5th edn. Reading, MA: Addison-Wesley.

Spence, M. 1981: The learning curve and competition. *Bell Journal of Economics*, 12, 49–70.

Stokey, N. 1986: The dynamics of industry-wide learning. In W. Heller, R. Starr and D. Starrett (eds.), *Equilibrium Analysis: Essays in Honor of Kenneth J. Arrow*, vol. II. Cambridge: Cambridge University Press.

PART II

Protection of Domestic Industry and International Treaties

The Partial Equilibrium Analysis of International Trade

■ **6.0 INTRODUCTION**

Countries use trade policy as a means of achieving a variety of goals. Developed countries often protect agriculture as a way of providing benefits to agricultural producers or achieving environmental objectives. Developing countries often tax agriculture as a way of obtaining revenue or providing low-cost food to consumers.

Partial equilibrium trade analysis is useful for analyzing the welfare effects of agricultural trade policies. As opposed to general equilibrium trade analysis, which provides a broad picture of the impact of trade policy on the economy as a whole, partial equilibrium analysis focuses on the direct impacts of a trade policy and provides useful information for policy-makers. This information includes impacts on domestic and world prices, production, consumption, trade, and welfare with respect to the commodity in question. Although there are other indirect impacts that partial equilibrium trade analysis does not account for, the information that it provides is useful to policy-makers as they formulate agricultural and trade policies.

This chapter presents the foundation for partial equilibrium trade analysis through the derivation of the import demand and export supply equations. By combining these two concepts, the determination of a trade equilibrium in a partial equilibrium framework is introduced. This chapter also identifies and discusses the divergence in domestic prices due to transportation and other costs.

■ **6.1 IMPORT DEMAND AND DEMAND ELASTICITIES**

THE IMPORT DEMAND FUNCTION

Import demand for a commodity occurs where the domestic quantity of the good demanded is greater than the domestic quantity of the good supplied at a given price.

Thus, import demand is referred to as excess demand. The import demand function is defined as

$$Q_m(P) = Q_d(P) - Q_s(P)$$ (6.1)

where $Q_m(P)$ is the quantity of the good imported at price P, $Q_d(P)$ is the quantity of the good demanded at price P, and $Q_s(P)$ is the quantity of the good supplied at price P. Because domestic demand and supply of a commodity depend upon price, import demand also has a functional relationship with price. If the price increases, domestic production of the commodity increases, while domestic demand for the commodity decreases. This results in a decrease in imports of the commodity. Conversely, if the price decreases, domestic production of the commodity decreases, while domestic consumption increases. This results in an increase in imports. Given this, the import demand for a good is inversely related to price.

Import demand for a commodity can be derived from the domestic demand and supply of a commodity, as shown in Figure 6.1. Figure 6.1(a) shows the domestic demand and supply schedules of a good. The equilibrium price is obtained at point E, where domestic demand for the good intersects domestic supply at a price of $50 per unit. At this price, the quantity demanded (30 units) equals the quantity supplied (30 units), indicating that the quantity imported equals zero. If the price of the commodity declines from $50 to $30, domestic supply declines from 30 units to 20 units, and domestic demand increases from 30 units to 40 units. As a result, at a price of $30, this country imports 20 units of the commodity, which is equal to a′b′ in Figure 6.1(b). Similarly, when the price of the commodity declines further from $30 to $20, domestic supply decreases from 20 units to 15 units, and domestic consumption increases from 40 units to 45 units. This results in the import of 30 units of the commodity, which is equal to c′d′ in Figure 6.1(b). Connecting the import points at their respective prices forms the import demand schedule, as shown in Figure 6.1(b). This downward slope indicates that the import demand for a good has an inverse relationship with the price of the good.

SOCIAL WELFARE ANALYSIS

All buyers must pay the market price, $50 per unit in the equilibrium as shown in Figure 6.1(a), but that is the price that only a marginal buyer is willing to pay. Other buyers, who are more eager for the product, are willing to pay higher prices, as indicated by the domestic demand curve above the market price of $50. However, they pay the same market price. The difference between what consumers are willing to pay and the market price that they actually pay is known as consumer surplus. In Figure 6.1, consumer surplus is measured by the area of the triangular area I at market price, $50. When the market price decreases from $50 to $30, consumer surplus is measured by the triangular area A + B + I. As the market price decreases, consumer surplus increases, indicating that consumers are better off. With the decreased

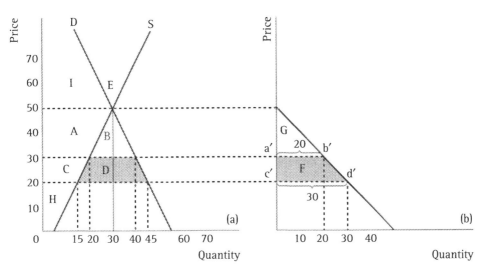

Figure 6.1 Derivation of the import demand curve.

market price, $30, the increase in consumers' surplus is area $A + B$ (the difference between the triangular areas I and $A + B + I$).

Next, consider the supply curve in Figure 6.1(a). Points on the supply curve show the quantities that sellers are willing to supply at various prices. All sellers receive the market price, $50, in the equilibrium condition and supply 30 units, but some efficient suppliers are willing to supply at lower prices, as indicated by the domestic supply curve below the market price of $50. The difference between the price that sellers could receive and the market price that they actually receive is known as producer surplus. In Figure 6.1(a), producer surplus is measured by triangular area $A + C + H$ at a price of $50. When the market price decreases from $50 to $30, the producer surplus is measured by triangular area $C + H$, indicating that producer surplus decreases. This decrease in producers' surplus is the area represented by A.

Hence, a decrease in price from $50 to $30 results in a gain to consumers of area $A + B$ and a loss to producers of area A. Thus, there is a net increase in social benefits of triangular area B, which is equal to triangular area G in Figure 6.1(b). Similarly, as the price decreases further from $30 to $20, the net increase in social benefits is equal to triangular area $B + D$, which is equal to triangular area $G + F$ in Figure 6.1(b).

PRICE ELASTICITY OF IMPORT DEMAND

The relationship between the quantity of a good imported and price of the good can be shown using the concept of price elasticity. The price elasticity of import demand is defined as the percentage change in the quantity imported resulting from a 1% change in the price of the imported commodity. This is mathematically expressed as follows:

$$e_m = \frac{\%\Delta Q_m}{\%\Delta P} = \frac{\Delta Q_m / Q_m}{\Delta P / P} = \frac{\Delta Q_m}{\Delta P} \cdot \frac{P}{Q_m} \tag{6.2}$$

where e_m is the price elasticity of import demand, ΔQ_m represents the change in imports of a good, ΔP represents the change in the price of the good, P is the price of the good, and Q_m is the quantity of the good imported.

Assume that the quantity of wheat imported by Japan is 4.5 million bushels, at an import price of $4.00 per bushel. Japanese imports of wheat increase to 5.1 million bushels when the price decreases to $3.50 per bushel. In this example, the Japanese price elasticity of import demand for wheat is calculated as follows:

$$e_m = \frac{0.6}{-0.5} \times \frac{4.0}{4.5} = -1.06$$

The price elasticity of import demand (e_m) is -1.06, indicating that a 1% increase in price results in a 1.06% decrease in quantity imported.

Alternatively, the price elasticity of import demand (e_m) can be calculated using the domestic demand elasticity (e_d) and supply elasticity (e_s) in equation (6.3) below. The derivation of this equation is shown in Appendix 6.1. The price elasticity of import demand is calculated as follows:

$$e_m = e_d \left(\frac{Q_d}{Q_m} \right) - e_s \left(\frac{Q_s}{Q_m} \right) \tag{6.3}$$

where e_d is the price elasticity of domestic demand, e_s is the price elasticity of domestic supply, Q_s is the quantity of the good supplied, and Q_d is the quantity demanded.

Assume that a country's price elasticity of domestic demand (e_d) for wheat is -0.5 and that the price elasticity of domestic supply (e_s) is 1.0 in Japan. Also assume that domestic demand for wheat (Q_d) is 6 million bushels and that the domestic supply of the product (Q_s) is 1.5 million bushels, at a market price of $4.00 per bushel. Then import demand is the difference between Q_d and Q_s. In this case, the price elasticity of import demand for wheat (e_m) is calculated, using equation (6.3), as follows:

$$e_m = -0.5(6.0/4.5) - 1.0(1.5/4.5) = -1.0$$

The price elasticity of import demand for wheat (e_m) is -1.0, indicating that a 1% increase in the price of wheat results in a 1% decrease in the quantity of wheat imported.

■ 6.2 EXPORT SUPPLY AND ELASTICITIES

THE EXPORT SUPPLY FUNCTION

The export supply of a commodity occurs where the domestic supply of the commodity is larger than the domestic demand at a given price. Thus, export supply is also known as excess supply. The export supply function is defined as follows:

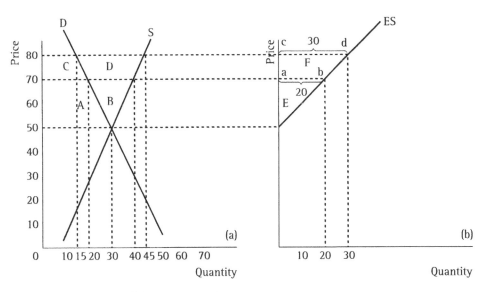

Figure 6.2 Derivation of the export supply curve.

$$Q_x(P) = Q_s(P) - Q_d(P) \tag{6.4}$$

where $Q_x(P)$ is the export supply of the good at price P. The other variables were defined previously. Since the domestic demand and supply of a good are functions of price, export supply is also a function of price. This relationship is shown in Figure 6.2.

As the price increases from $50 to $70, domestic supply increases from 30 to 40 units, while domestic demand decreases from 30 to 20 units. As a result, this country exports 20 units of the commodity, equivalent to ab in Figure 6.2(b). As the price increases further from $70 to $80, domestic supply is 45 units and domestic demand is 15 units, resulting in the export of 30 units. This export quantity is represented as cd in Figure 6.2(b). Connecting the export points at their respective prices forms the export supply schedule, as shown in Figure 6.2(b). Export supply has an upward slope, indicating that exporters increase the quantity of goods supplied as the price increases.

SOCIAL WELFARE ANALYSIS

As the price increases from $50 to $70, the increase in producer surplus is $A + B$ and the decrease in consumer surplus is A, as shown in Figure 6.2(a). The increased price results in an increase in social benefits by triangular area B, which is equal to triangular area E in Figure 6.2(b). Similarly, as the price increases further from $70 to $80, the net increase in producer surplus is area $C + D$, and the net decrease in consumer surplus is area C, indicating that the net increase in social benefits is equal to triangular area D, which is equal to area F.

PRICE ELASTICITY OF EXPORT SUPPLY

Price elasticity of export supply is defined as the percentage change in the quantity of a commodity supplied resulting from a 1% change in the price of the commodity. The price elasticity is expressed mathematically as follows:

$$e_x = \frac{\%\Delta Q_x}{\%\Delta P} = \frac{\Delta Q_x/Q_x}{\Delta P/P} = \frac{\Delta Q_x}{\Delta P} \times \frac{P}{Q_x} \tag{6.5}$$

where e_x is the price elasticity of export supply, ΔQ_x represents the change in exports of a good, ΔP represents the change in the price of the good, P is the price, and Q_x is the quantity of the exported good.

Assume that the quantity of wheat that the United States exports to Japan is 4.5 million bushels, at an export price of $4.00 per bushel. A price increase from $4.00 to $4.50 causes US exports to Japan to increase from 4.5 million bushels to 5.1 million bushels. In this example, the price elasticity of US export supply is calculated as follows:

$$e_x = \frac{0.6}{0.5} \times \frac{4.0}{4.5} = 1.06$$

The elasticity of 1.06 indicates that a 1% change in the price of wheat results in 1.06% change in US exports of wheat to Japan.

Alternatively, the price elasticity of export supply can be calculated from the domestic demand and supply elasticities. The derivation of the equation is shown in Appendix 6.2. The price elasticity of export supply is calculated as follows:

$$e_x = e_s \left(\frac{Q_s}{Q_x} \right) - e_d \left(\frac{Q_d}{Q_x} \right) \tag{6.6}$$

where e_x is the price elasticity of export supply, e_s is the price elasticity of domestic supply, e_d is the price elasticity of domestic demand, Q_d is the quantity of the good demanded, and Q_s is the quantity of the good supplied.

Assume that the price elasticity of domestic demand for wheat (e_d) is -1.0 and that the price elasticity of domestic supply (e_s) is 0.5 in the USA. Also assume that the domestic supply of wheat (Q_s) is 6.0 million bushels and that domestic demand (Q_d) is 1.5 million bushels. The exportable surplus is 4.5 million bushels, which is the difference between Q_s and Q_d. In this example, the price elasticity of export supply (e_x) is calculated, using equation (6.6), as follows:

$$e_x = 0.5(6.0/4.5) - (-1.0)(1.5/4.5) = 1.0$$

The price elasticity of export supply (e_x) is 1.0, indicating that a 1% increase in prices results in a 1.0% increase in quantity of this good exported.

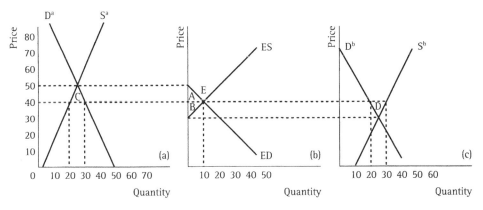

Figure 6.3 International equilibrium between countries A and B: (a) country A; (b) international market; (c) country B.

6.3 INTERNATIONAL EQUILIBRIUM

Figure 6.3(a) shows domestic demand and supply schedules in an importing country (A), and Figure 6.3(c) shows domestic demand and supply schedules in an exporting country (B). Import demand (ED) and export supply (ES) schedules are derived from these domestic demand and supply schedules as shown in the previous section.

Before opening international trade, the equilibrium price in country A is $50 per unit with a quantity supplied and demanded of 25 units. In country B, the equilibrium price is $30 and the quantity supplied and demanded is 25 units. When trade is allowed, the international equilibrium occurs at point E where the export supply intersects the import demand in Figure 6.3(b). The equilibrium price is equal to $40, which corresponds with the equilibrium volume of 10 units traded in the international market. This trade volume is equal to country A's imports and country B's exports.

The net increase in the total social welfare is the sum of triangular areas A and B. The upper triangle (area A) is the net increase in the social welfare of consumers in the importing country (area C in Figure 6.3(a)). The lower triangle (area B) is the net increase in the social welfare of producers in the exporting country (area D in Figure 6.3(c)).

6.4 INTERNATIONAL EQUILIBRIUM IN SMALL IMPORTING AND EXPORTING COUNTRIES

A small importing country faces a perfectly elastic export supply function, because the country's imports are not large enough to influence the world price. This trade relationship is shown in Figure 6.4. Before opening international trade, this country's domestic supply equals its domestic demand at the equilibrium price of $40. The quantity of the good supplied and demanded is 30 units.

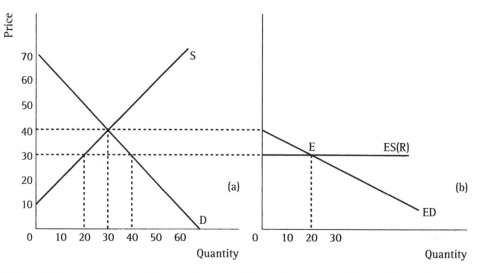

Figure 6.4 International equilibrium for a small importing country: (a) small importing country; (b) international market.

Since the world price is exogenous for the small importing country, this indicates that the small importer is a price taker. The world equilibrium price is obtained at point E where the country's ED schedule intersects the perfectly elastic ES schedule. The export supply schedule the country faces is parallel to the x-axis. The equilibrium world price is $30 per unit and the quantity traded in the world market is 20 units. The country's domestic price decreases from $40 to $30 after starting to import the good from the rest of the world. At the domestic price of $30, domestic demand is 40 units and domestic supply is 20 units. This country imports 20 units at the world price of $30 per unit.

Similarly, the small exporting country cannot influence the world market (see Figure 6.5). This means that the country is a price taker. The import demand schedule the country faces is parallel to the x-axis. Before opening international trade, domestic demand is equal to domestic supply at market price, $30. The world equilibrium price is determined at point E where the export supply schedule (ES) intersects the import demand schedule (ED). The world equilibrium price is $40 per unit and the quantity traded in the world market is 20 units. At the price of $40 per unit, the country's domestic supply is 40 units and domestic demand is 20 units. This country exports 20 units to the world market at the market price of $40 per unit.

■ 6.5 EFFECTS OF TRANSFER COST

Consider the impact of transportation costs in shipping a commodity from country B to country A. Figure 6.6(b) shows that the world equilibrium price without transportation

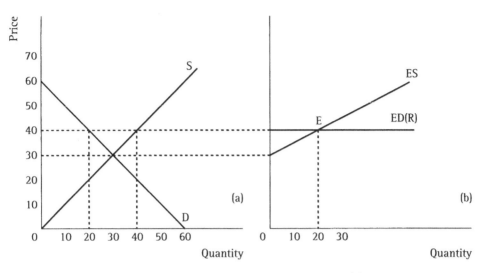

Figure 6.5 International equilibrium for a small exporting country: (a) small importing country; (b) international market.

costs is $40 and the quantity traded is 10 units. Transportation costs are represented by distance ab, which is equal to $10 per unit in Figure 6.6. Given this, the price in the importing country increases from $40 to $45, and the price in the exporting country decreases from $40 to $35. The price difference between the two countries, which is $10 per unit, is equal to the transportation costs in a free market system. As a result of an increase in price in the importing country and a decrease in price in the exporting country, the trade volume declines from 10 units to five units. This trade volume is equal to country A's imports (five units) and country B's exports (five units). The decrease in the price in the exporting country ($5 per unit) is the portion of transportation costs that producers pay in the exporting countries. The increase in price in the importing country ($5 per unit) is the portion that consumers pay in the importing countries. This means that the burden of transportation costs is shared between the two countries, depending upon the price elasticities of export supply and import demand. The incidence of transportation costs can be calculated as follows:

$$s = \frac{1}{|e_m/e_x| + 1} \tag{6.7}$$

where s is the ratio of transportation costs paid by consumers in an importing country to the total transportation cost, e_m is the price elasticity of import demand, and e_x is the price elasticity of export supply. The derivation of this formula is presented in Appendix 6.3.

If the price elasticity of import demand for a commodity (e_m) is equal to that of export supply (e_x) in the absolute value, then s is equal to 0.5, indicating that the

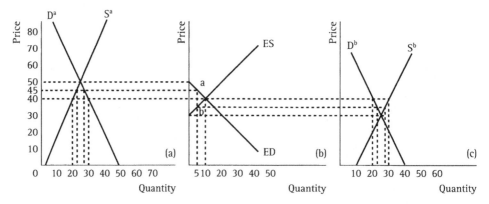

Figure 6.6 International equilibrium between countries A and B with transportation costs: (a) country A; (b) international market; (c) country B.

transportation cost is shared evenly between the two countries. In Figure 6.6, it is assumed that the price elasticity of import demand (e_m) is equal to that of export supply (e_x). When a transportation cost of $10 per unit is introduced, the price in the importing country increases from $40 to $45, while the price in the exporting country decreases from $40 to $35, indicating that the transportation cost is shared evenly between the two countries.

If the price elasticity of import demand (e_m) is larger than that of export supply (e_x) in absolute value, as shown in Figure 6.7, then s is smaller than 0.5. This indicates that the portion of transportation costs that consumers pay is smaller than what producers pay. In Figure 6.7, the world equilibrium price is $50 per unit and the quantity traded is 40 units with no transportation cost.

When a transportation cost of $20 per unit is introduced, the price in the importing country increases from $50 to $55, while the price in the exporting country decreases from $50 to $35. This indicates that the incidence of the transportation cost paid by the exporting country ($15 per unit) is much larger than that paid by the importing country ($5 per unit.) Conversely, if the price elasticity of import demand (e_m) is less elastic than that of export supply (e_x) in absolute value, as shown in Figure 6.8, then s is greater than 0.5. This implies that the portion of transportation costs that consumers pay is larger than what producers pay. In Figure 6.8, when a transportation cost of $20 per unit is introduced, the price in the importing country increases from $50 to $65, while the price in the exporting country decreases from $50 to $45. The incidence given to consumers in the importing country is $15 per unit and that given to producers in the exporting country is $5 per unit, indicating that a greater portion of the transportation cost is paid by consumers in the importing country.

For a small importing country facing a perfectly elastic export supply schedule, s is equal to 1, indicating that transportation costs are completely borne by consumers in the importing country. On the other hand, for a small exporting country facing a perfectly elastic import demand, s is equal to zero, indicating that producers in the exporting country bear all transportation costs.

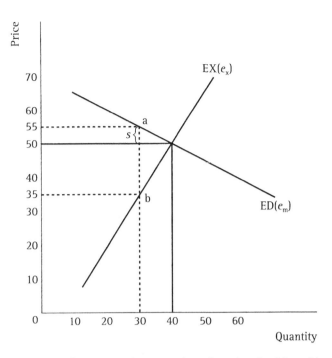

Figure 6.7 The incidence of transportation cost when the price elasticity of import demand is more elastic than that of export supply.

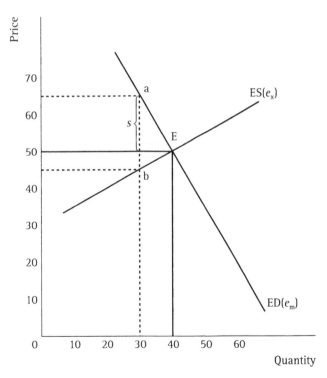

Figure 6.8 The incidence of transportation cost when the price elasticity of import demand is less elastic than that of export supply.

BOX 6.1 TRANSPORTATION COSTS IN INTERNATIONAL TRADE

Partial equilibrium trade analysis often assumes zero transportation costs. This simplifies the issue and allows researchers to determine the direction of price movements when considering the impact of various trade policies. Outcomes using this assumption typically show that differences in prices among countries are solely the result of agricultural policies.

In practice, the cost of transportation, which includes insurance, handling and freight, results in a price wedge between importing and exporting market prices. An example of this involves the price differential between the free on board (F.O.B.) and cost, insurance, and freight (C.I.F.) price quotes. Assume that rice is being exported from Thailand to Bangladesh by vessel. In the case of exporting the product, the F.O.B. price includes the price of the product plus the cost of moving the rice to the dock and loading it on the ship. The C.I.F. price is the price of the product plus freight and any other transportation-related costs. Export revenue is closely associated with the F.O.B. price and the cost to importers is more closely related to the C.I.F. quote. On this basis, it is clear that there is a price wedge due to the cost of transportation.

An example of this price differential can be seen by examining the wholesale prices for milled rice in Bangladesh and Thailand. As shown in Table 6.1, the wholesale price of milled rice in the exporting country, Thailand, is $281.63 per ton. For the importing country, Bangladesh, the wholesale price of milled rice is $339.03 per ton. A $57.40 price wedge exists between the two countries. A portion of the price wedge could be due to agricultural protection. The remainder is due to transportation costs.

Suppose that there are no barriers to trade and transportation costs are zero. If a price differential exists, traders will perform arbitrage. That is, they will buy the product in the low-priced market and ship it at no cost to the high-priced market, selling the product for a profit. Increased demand in the export market and increased supply in the import market narrow the price differential. Eventually the price wedge will disappear, eliminating any incentive for further trade. However, transportation costs result in price differentials between trading partners that will continue even if free trade is achieved.

Table 6.1 A comparison of the wholesale milled rice prices in Bangladesh and Thailand, 1997

	Domestic price (local currency per ton)	Exchange rate (local currency per dollar)	Domestic price (dollars per ton)
Bangladesh	14,880	43.89	339.03
Thailand	8,832	31.36	281.63

Source: *World Rice Statistics* (2003), International Rice Research Institute, data on the internet (http://www.irri.org/science/ricestat/index.asp, document accessed July 10, 2003) and *International Financial Statistics* (March 2003), International Monetary Fund, data on diskette, Washington, DC

SUMMARY

1 The import demand for a commodity occurs when domestic demand for the commodity is greater than domestic supply of the commodity at a given price. The import demand schedule in an importing country can be derived from domestic demand and supply schedules of the commodity in the country and has an inverse relationship with price.

2 Consumers in the importing country are better off through international trade because opening trade for a good lowers the domestic price of the good and results in an increase in consumer surpluses, which is greater than a decrease in producer surpluses in that country.

3 The export supply of a commodity occurs when the domestic supply of the commodity is larger than the domestic demand at a given price. The export supply schedule in an exporting country can be derived from the domestic demand and supply schedules for a commodity in the country and has a positive functional relationship with price.

4 Producers in the exporting country are better off through international trade because exports of the commodity increase the domestic price of the commodity and results in an increase in producer surplus, which is large enough to offset a decrease in consumer surplus.

5 A small importing country faces a perfectly elastic export supply function, assuming that the country's trade volume is not large enough to influence the world market. In this case, the importing country is a price taker. On the other hand, a small exporting country faces a perfectly elastic import demand function, assuming that the country does not influence the world market. This exporting country is also a price taker.

6 The incidence of transportation costs is shared between producers in exporting countries and consumers in importing countries, depending upon the magnitudes of the import demand and export supply elasticities. If the price elasticity of import demand is more elastic than that of export supply, producers in the exporting country bear a larger portion of the transport costs than consumers in the importing country. Conversely, if the price elasticity of export supply is more elastic than that of import demand, consumers in the importing country bear a larger portion of the transport costs. If price elasticities of import demand and export supply are the same in absolute terms, the transportation costs are shared evenly between the two countries.

KEY CONCEPTS

Consumer surplus – The difference between what consumers would have been willing to pay and the market price that they actually pay for a good.

Export supply – Quantities of a good exported at alternative export prices.

Export supply elasticity – The percentage change in the quantity of a good exported resulting from a 1% change in the export price of the good.

Import demand – Quantities of a good imported at alternative import prices.

Import demand elasticity – The percentage change in the quantity of a good imported resulting from a 1% change in the import price of the good.

Incidence of transport costs – The burden of shipping cost paid by importing countries and that paid by the exporting country.

Producer surplus – The difference between the price that the sellers would have been willing to receive and the market price that they actually receive for a good.

Transportation costs – The costs that occur in shipping a good between two countries.

QUESTIONS AND TASKS FOR REVIEW

1   How does a partial equilibrium analysis differ from the general equilibrium approach?
2   What is meant by export supply of a good in a country? How can it be driven from domestic demand and supply schedules?
3   What is meant by import demand for a good in a country? How can it be driven from domestic demand and supply schedules?
4   What is meant by consumer surplus, and how can it be measured with a given demand schedule and market price?
5   What is meant by producer surplus, and how can it be measured with a given supply schedule and market price?
6   What do price elasticities of export supply and import demand mean, and how can they be calculated?
7   What is the relationship between price elasticity of import demand and price elasticities of domestic demand and supply in an importing country?
8   What is the relationship between price elasticity of export supply and price elasticities of domestic demand and supply in an exporting country?
9   Why are consumers in an importing country better off when it imports a commodity from other countries? Explain using the concepts of consumer and producer surplus.
10 Are producers in an exporting country better off when it starts to export a commodity to other countries? Explain using the concepts of consumer and producer surplus.
11 Explain the incidence of transport costs in shipping a commodity from exporting countries to importing countries.

SELECTED BIBLIOGRAPHY

Grennes, T. 1984: *International Economics*. Englewood Cliffs, NJ: Prentice Hall.
Houck, J. P. 1992: *Elements of Agricultural Trade Policies*. Prospect Heights, IL: Waveland Press.
McCalla, A. and Josling, T. 1985: *Agricultural Policies and World Markets*. New York: Macmillan Press.

APPENDIX 6.1 DERIVATION OF PRICE ELASTICITY OF IMPORT DEMAND

Import demand is defined as the difference between domestic demand and supply, as follows:

$$Q_m = Q_d - Q_s \tag{A6.1-1}$$

Combining equations (A6.1-1) and (6.2) yields

$$e_m = \frac{\Delta(Q_d - Q_s)}{\Delta P} \times \frac{P}{Q_m} \tag{A6.1-2}$$

Equation (A6.1-2) can be rewritten as follows:

$$e_m = \frac{\Delta Q_d}{\Delta P} \times \frac{P}{Q_m} - \frac{\Delta Q_s}{\Delta P} \times \frac{P}{Q_m} \tag{A6.1-3}$$

Multiply the first term of equation (A6.1-3) by (Q_d/Q_d) and the second term by (Q_s/Q_s), as follows:

$$e_m = \frac{\Delta Q_d}{\Delta P} \times \frac{P}{Q_m}\left(\frac{Q_d}{Q_d}\right) - \frac{\Delta Q_s}{\Delta P} \times \frac{P}{Q_m}\left(\frac{Q_s}{Q_s}\right) \qquad \text{(A6.1-4)}$$

Equation (A6.1-4) can be summarized as

$$e_m = \frac{\Delta Q_d}{\Delta P} \times \frac{P}{Q_d}\left(\frac{Q_d}{Q_m}\right) - \frac{\Delta Q_s}{\Delta P} \times \frac{P}{Q_s}\left(\frac{Q_s}{Q_m}\right) \qquad \text{(A6.1-5)}$$

Since

$$\frac{\Delta Q_d}{\Delta P} \times \frac{P}{Q_d} = e_d \quad \text{and} \quad \frac{\Delta Q_s}{\Delta P} \times \frac{P}{Q_s} = e_s$$

Equation (A6.1-5) can be rewritten as

$$e_m = e_d\left(\frac{Q_d}{Q_m}\right) - e_s\left(\frac{Q_s}{Q_m}\right) \qquad \text{(A6.1-6)}$$

APPENDIX 6.2 DERIVATION OF PRICE ELASTICITY OF EXPORT SUPPLY

Export supply is defined as the difference between domestic demand and supply, as follows:

$$Q_x = Q_s - Q_d \qquad \text{(A6.2-1)}$$

Combining equations (A6.2-1) and (6.5) yields

$$e_x = \frac{\Delta(Q_s - Q_d)}{\Delta P} \times \frac{P}{Q_x} \qquad \text{(A6.2-2)}$$

Equation (A6.2-2) is rewritten as

$$e_x = \frac{\Delta Q_s}{\Delta P} \times \frac{P}{Q_x} - \frac{\Delta Q_d}{\Delta P} \times \frac{P}{Q_x} \qquad \text{(A6.2-3)}$$

Equation (A6.2-3) can be expressed as

$$e_x = \frac{\Delta Q_s}{\Delta P} \times \frac{P}{Q_x}\left(\frac{Q_s}{Q_s}\right) - \frac{\Delta Q_d}{\Delta P} \times \frac{P}{Q_x}\left(\frac{Q_d}{Q_d}\right) \qquad \text{(A6.2-4)}$$

Since

$$\frac{\Delta Q_s}{\Delta P} \times \frac{P}{Q_x} = e_s \quad \text{and} \quad \frac{\Delta Q_d}{\Delta P} \times \frac{P}{Q_d} = e_d$$

Equation (A6.2-4) can be written as

$$e_x = e_s\left(\frac{Q_s}{Q_x}\right) - e_d\left(\frac{Q_d}{Q_x}\right)$$

(A6.2-5)

APPENDIX 6.3 DERIVATION OF THE INCIDENCE OF TRANSPORTATION COSTS

The price elasticity of import demand is defined as

$$e_m = \frac{\Delta Q_m}{\Delta P} \times \frac{P}{Q_m}$$

(A6.3-1)

where e_m is the price elasticity of import demand, Q_m is the quantity imported, and P is the import price.

Assuming that s is a ratio of the transportation costs consumers in the importing countries pay to the total transportation cost per unit (t), the price elasticity of import demand can be rewritten as follows:

$$e_m = \frac{\Delta Q_m}{st} \times \frac{P}{Q_m}$$

(A6.3-2)

where t is the transportation cost per unit and $\Delta p = st$.

Equation (A6.3-2) can be rewritten as

$$\frac{\Delta Q_m}{Q_m} = e_m \frac{st}{p}$$

(A6.3-3)

Now consider the price elasticity of export supply. It is defined as follows:

$$e_x = \frac{\Delta Q_x}{\Delta P} \times \frac{P}{Q_x} = \frac{\Delta Q_x}{Q_x} \times \frac{P}{(1-s)t}$$

(A6.3-4)

where e_x is the price elasticity of export supply, Q_x is the quantity exported, and P is the export price. $(1-s)$ is defined as a ratio of the transportation cost paid by producers in exporting countries to the total transportation cost.

Equation (A6.3-4) can be rewritten as

$$\frac{\Delta Q_x}{Q_x} = e_x \frac{(1-s)t}{P}$$

(A6.3-5)

Combining equations (A6.3-3) and (A6.3-5) under an equilibrium condition where $Q_x = Q_m$ gives

$$e_x = e_m \frac{s}{1-s}$$

(A6.3-6)

Solving for s yields

$$s = \frac{1}{|e_m/e_x| + 1}$$

(A6.3-7)

Trade Restrictions: Tariffs

7.0 INTRODUCTION

Many nations support the idea that free trade based on comparative advantage improves global welfare. At least, they support this concept in principle and are actively involved in multilateral trade agreements (GATT and WTO) and regional trade agreements (NAFTA, FTAA, European Union (EU), etc.).

Despite this evidence, the spirit of free trade has been shaded by the imposition of trade restrictions. Many of these countries have identified sectors of their economy that warrant protection. In fact, protection for agricultural commodities is typically much greater than for industrial commodities. This protection is often achieved by limiting imports, either through quantitative restrictions or taxes.

This chapter examines why many countries protect their agricultural sector despite evidence that protection does not improve social utility. The effects of the most common trade barrier, the tariff, are also discussed. In addition, the concepts of nominal tariffs and effective protection are introduced, along with the idea of an optimal tariff.

7.1 ARGUMENTS FOR PROTECTION

The gains from free trade were discussed in Chapters 2, 3, and 4. These chapters demonstrated that free trade leads to a high productivity and income for producers and a higher level of social utility for consumers than no trade (autarky). Free trade enables each country to obtain a higher level of social utility through increases in production and consumption than can be obtained in autarky. Free trade optimizes utilization of the given resources in the production of goods through specialization.

In addition, free trade maximizes consumer utility, since prices of goods are lower under free trade. Free trade stimulates competition among nations. As a result, production becomes more efficient than in isolation. Under perfect competition, free trade meets the requirements of Pareto optimality. In other words, no one can be better off without making someone else worse off.

However, world trade is far from free. Most nations use various trade barriers to protect relatively inefficient industries. This is especially true for agriculture. The average tariff for agricultural goods (30%) is much higher than that for industrial goods (6%). All nations accept the importance of free trade to improve production efficiency and consumer utility, and to obtain benefits from free trade by increasing their exports. However, they are reluctant to open their markets to foreign goods. There have been international efforts to promote worldwide free trade. There have been eight rounds of multilateral trade negotiations since 1947 to reduce trade barriers for agricultural and industrial goods, and many regional and bilateral trade agreements to promote free trade in the regions. However, nations still use various trade barriers (tariff and nontariff) to protect their domestic industries. This significantly reduces the social utility of nations.

There have been many different justifications for protectionism. Some are fallacious arguments for protection, including (1) keeping money at home by reducing or eliminating imports, (2) protecting the home market for domestic producers, and (3) equalizing the price of imports and domestic goods (scientific tariff), allowing domestic producers to meet foreign competition. The first argument indicates that if a country imports, it gets the goods and the foreign exporter gets money, while the country gets both goods and money if the country consumes domestic goods. However, this argument is flawed because money, as a means of exchange, could be returned in payment for exports or as an investment. The second argument (home market) is also weak, because a policy that protects domestic markets for inefficient domestic producers results in inefficient utilization of productive resources. A nation must reduce the size of its inefficient industries to be competitive in the global market. If a nation were to impose tariffs to equalize the prices of imports and domestic prices, this would eliminate international competition – the basis for international trade.

More appealing arguments that have been put forward involve (1) creating jobs in domestic industries by reducing imports and (2) curing a deficit in the nation's balance of payments. Protection would temporarily reduce domestic unemployment and a trade deficit. However, the policy would cause greater unemployment and trade deficits abroad, resulting in retaliation from other nations. As a result, all nations would be worse off in the end. Domestic unemployment and trade deficits would be better corrected through appropriate trade, monetary, and fiscal policies rather than trade restrictions.

There are a few qualified arguments for protection. They are justified from either noneconomic or economic considerations. Some of the arguments are as follows:

1 *National security.* All nations need to maintain an adequate national defense. To maintain national defense, a nation needs to protect its strategic domestic industry, which produces military hardware. On the other hand, some exporting countries, which are capable of producing sophisticated military hardware, need to control exports of this military hardware (e.g., long-distance missiles, nuclear weapons, tanks, etc.) to avoid movements of these items to unfriendly, hostile countries.

The rationale for national security is also used to protect agricultural goods. Suppose that a country imports most of its food needs. When the country has a conflict with suppliers, or when shipping channels are disrupted by military conflicts in the region, it may not have an adequate supply of food. To avoid this possibility, most countries want to produce some minimum level of agricultural commodities to meet domestic demand, even though their agricultural sector may not be competitive. This argument has been used by many countries, including Korea and Japan. These two countries experienced food shortages during World War II, because they were not able to import food during that period. In addition, some countries have used trade as a diplomatic weapon to disrupt the food supply in targeted countries. A good example of this argument is the US grain embargo against the Soviet Union (the USSR) following the USSR's invasion of Afghanistan. The US grain embargo disrupted the USSR's food supply and demand system for a few years.

2 *Unfair trade.* When a country subsidizes its agriculture and tries to export these commodities, the country gains an advantage in competing with other countries producing the same commodities. In this case, the other countries try to protect their industry from the subsidized foreign competitors. In addition, businesses, farmers and producers often argue that foreign governments play by a different set of rules than the home government, giving foreign firms unfair advantages. They insist that import restrictions should be enacted to offset these foreign advantages. The US government imposes a countervailing duty on imported commodities that are unfairly subsidized by foreign governments.

Another example of unfair trade is dumping. Dumping is defined as selling goods in an importing country at prices below those prevailing in the exporting country or selling goods below production costs of the goods. If the importing country has no domestic industry that is competing with the dumped products, buyers in the importing country enjoy a continuous benefit from lower prices of foreign goods. On the other hand, if the import country does have a domestic industry that is competing with the foreign products, the dumping will be harmful for domestic producers. When dumping is either sporadic – intended to harass a competing domestic industry – or predatory, it becomes undesirable. In this case, protection for the domestic industry is necessary.

3 *Infant industry.* One of the more commonly accepted cases for the protection of a domestic industry is the infant-industry argument. A country may have a potential comparative advantage in a commodity, but because of a lack of technology or experience in producing the commodity, firms in the country cannot compete with more established foreign firms. In this case, temporary trade protection is justified

to establish the domestic industry during its infancy. This argument is more justified for developing countries than for industrial countries. However, domestic subsidies may be more efficient than trade protection in helping the domestic industry to be more competitive. This argument is more applicable to agricultural processing and input industries as opposed to agricultural commodities. In some developing countries, the food processing industry is not competitive, mainly because of a lack of processing technology and capital, even though the countries have an abundance of primary goods for processing. In this case, the industry could be given an opportunity to develop its processing industry to compete with more developed countries.

■ 7.2 TARIFFS

A tariff (or customs duty) is the tax that a government imposes on commodities as they cross the national border. Tariffs have been used extensively to protect the domestic economy from foreign competition. A tariff imposed on imported goods is an import tariff. On the other hand, a government can impose tariffs on the country's exports of a commodity to generate tax revenue and/or to control export supply. A tariff imposed on exported goods is an export tariff.

A customs area is a geographical region within which commodities can move freely without restrictions. A customs area generally coincides with the national boundary. When a customs area includes more than one nation, it is a customs union.

In terms of their applications, tariffs are categorized into *ad valorem*, specific, and compound. The application methods of the different tariffs are as follows:

1 An *ad valorem* tariff is a fixed percentage of the price of a good. The tariff is calculated as follows: $t = \alpha p$, where α is a fixed tariff rate in percentage terms, and p is the price of the good.
2 A specific tariff is a fixed amount of money per physical unit of the good. A specific tariff is calculated as follows: $t = c$, where c is a fixed amount of money per unit of the good.
3 A compound tariff is a combination of *ad valorem* and specific tariffs. A compound tariff is calculated as follows: $t = c + \alpha p$.

An *ad valorem* tariff depends upon the price or value of a good, while a specific tariff is fixed regardless of the price or value of the good. Thus, a specific tariff provides less protection as prices of goods increase and provides stronger protection for cheaper goods. On the other hand, an *ad valorem* tariff provides a constant percentage protection rate for all goods as prices change. The *ad valorem* tariff also provides a constant percentage protection rate during inflationary periods, while the specific tariff decreases its protection rate under inflation.

In terms of application, the specific tariff is more simple than the *ad valorem* tariff. The *ad valorem* tariff is generally used for raw materials for which value can be evaluated easily. The specific tariff is generally used for manufactured goods, because the values of manufactured goods are not easily determined.

The application of tariffs is based on either the F.O.B. (free on board) price or the C.I.F. (cost, insurance, freight) price. The F.O.B. price is the price of a good at the port in the exporting country. This price includes loading charges. The C.I.F. price is the price of a good at the port in the importing country. The C.I.F. price includes transportation costs, insurance, and loading and unloading charges. Since C.I.F. prices are higher than F.O.B. prices, *ad valorem* and compound tariffs on C.I.F. prices differ from those on F.O.B. prices, even if the tariff rates are the same.

■ 7.3 THE INCIDENCE OF TARIFFS

Tariffs imposed by an importing country raise the consumer price of the commodity in the importing country and decrease the price of the commodity on the world market. For example, if the US government imposes an import tariff on coffee, it raises the price of coffee in the United States. This decreases the quantity of coffee demanded in the USA. The reduced US import demand reduces world demand for coffee. If the US imports of coffee are relatively large, the decreased demand for coffee in the world market lowers the world price of coffee. This implies that the world price of coffee declines because of the import tariff imposed by the USA. On the other hand, if the US imports of coffee are small, its impact on the world price is insignificant.

In Figure 7.1, ES is the export supply schedule of a commodity, and ED is the import demand schedule for the commodity. An equilibrium price, $50 per unit, is obtained when ES intersects ED at point E. In this case, the equilibrium quantity traded is 30 units. Assume that this country imposes a specific tax of $20 per unit. The vertical distance ab between ES and ED represents the importing country's specific import tariff imposed on each unit of the commodity ($20 per unit). As a result of the tariff, the price of the commodity in the importing country increases from $50 to $60, and the price in the exporting country declines from $50 to $40. The total trade volume decreases from 30 units to 20 units because of changes in prices in the importing and exporting countries. Since the importing country imposes a specific tariff, $20 per unit of the commodity, the total tariff revenue obtained by the importing country is area A + B, which is equal to $400 ($20 × 20 units).

The total tariff revenue is divided into two parts: one is paid by consumers in the importing country (area A) and the other is paid by producers in the exporting country (area B). The area A represents the importing country's tariff revenue paid by consumers in the importing country. This area also represents an income transfer from consumers to the government of the importing country. The area B represents the importing country's tariff revenue paid by producers in the exporting country. This implies that the burden of the tariff is shared by producers in the exporting country and consumers in the importing country.

The incidence of tariffs borne by importing and exporting countries can be calculated by using the formula introduced in Chapter 6. The fraction of the tariff paid by consumers in the importing country is calculated as

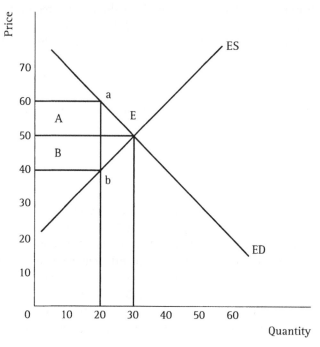

Figure 7.1 The impacts of a specific tariff on prices in importing and exporting countries.

$$s = \frac{1}{|e_m/e_x| + 1} \tag{7.1}$$

where e_m is the price elasticity of import demand and e_x is the price elasticity of export supply.

If the price elasticity of import demand is equal to that of export supply in absolute terms, as shown in Figure 7.1, the value of s in equation (7.1) is 0.5, indicating that the burden of the import tariff imposed by the importing country is shared evenly between consumers in the importing country and producers in the exporting country. The price increase ($10 per unit) in the importing country is equal to the price decrease ($10 per unit) in the exporting country when an import tariff of $20 per unit is imposed on a commodity. The tariff revenue paid by consumers in the importing country (area A) is, therefore, equal to that paid by producers in the exporting country (area B).

If price elasticity of export supply is less elastic than that of import demand, the value of s in equation (7.1) is smaller than 0.5, indicating that the burden of the import tariff borne by consumers in the importing country is smaller than that borne by producers in the exporting country. In Figure 7.2, the slope of ES is steeper than that of ED, indicating that ES is less elastic. If an importing country imposes an import tariff of $20 per unit, the price increases from $50 to $55 ($5 per unit) in the importing country, while the price decreases from $50 to $35 ($15 per unit) in the exporting country.

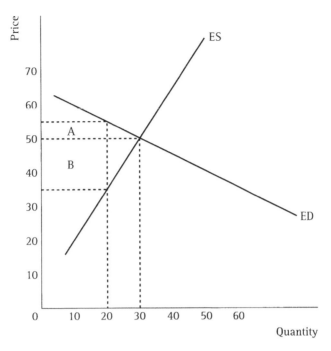

Figure 7.2 The incidence of a tariff if the price elasticity of export supply is less elastic than that of import demand.

The total tariff revenue collected by the importing country is $400 ($20 × 20 units), which is equal to area A + B. Since the total quantity traded in the world market decreases from 30 units to 20 units, the tariff revenue paid by consumers in the importing country (area A) is $100 ($5 × 20 units) and that paid by producers in the exporting country (area B) is $300 ($15 × 20 units).

If price elasticity of export supply is more elastic than that of import demand as shown in Figure 7.3, the value of s in equation (7.1) is larger than 0.5, indicating that the burden of the import tariff borne by consumers in the importing country is larger than that borne by producers in the exporting country. In Figure 7.3, the slope of ED is steeper than that of ES, indicating that ED is less elastic. The portion of the price increase ($15 per unit) in the importing country is larger than the price decrease ($5 per unit) in the exporting country. The quantity traded in the world market decreases from 30 units to 20 units. The total tariff revenue collected by the importing country is $400 ($20 × 20 units), which is equal to area A + B. The tariff revenue paid by consumers in the importing country (area A) is $300 ($15 × 20) and that paid by producers in the exporting country (area B) is $100 ($5 × 20 units).

For a small importing country, its import demand schedule faces a perfectly elastic export supply. In this case, the value of s in equation (7.1) is 1.0, indicating that the burden of the tariff is borne completely by consumers in the importing country. The

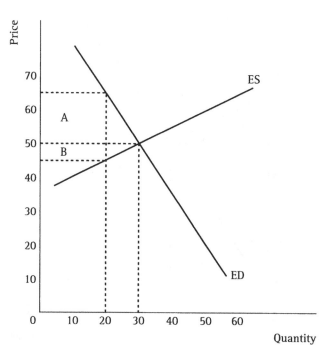

Figure 7.3 The incidence of a tariff if the price elasticity of export supply is more elastic than that of import demand.

tariff imposed by an importing country raises the price from $50 to $70 in the importing country, but does not change the world price, as shown in Figure 7.4.

Total tariff revenue collected by the importing country (area A) is $400 ($20 × 20 units), which is paid by consumers in the importing country.

Consider a small exporting country in which the export supply schedule faces a perfectly elastic import demand. In this case, if the exporting country imposes an export duty, the value of s in equation (7.1) approaches zero, indicating that the burden of the tariff imposed by the exporting country is borne completely by producers in the exporting country, as shown in Figure 7.5. The tariff decreases the domestic price from $50 to $30 in the exporting country, but does not change prices in the world market. As a result, the quantity traded decreases from 30 units to 20 units. The total tariff revenue collected by the exporting country (area B) is $400 ($20 × 20 units), all of which is paid by producers in the exporting country.

The incidence of tariffs also depends upon the characteristics of a good. If a good is a luxury good, import demand for the good could be more elastic compared to the supply of the good. Therefore, tariffs imposed on the luxury good are borne more by producers in the exporting country than by consumers in the importing country. On the other hand, if a tariff is imposed on a necessity, which has inelastic import demand, the burden of the tariff is typically borne more by consumers in the importing country than by producers in the exporting country.

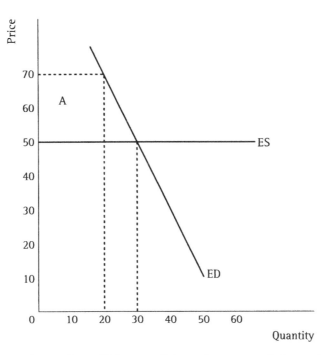

Figure 7.4 The incidence of a tariff for a small importing country facing perfectly elastic export supply.

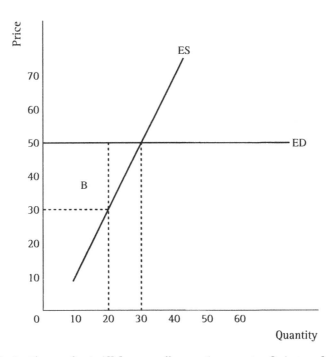

Figure 7.5 The incidence of a tariff for a small exporting country facing perfectly elastic import demand.

■ 7.4 WELFARE IMPLICATION OF TARIFFS

To simplify the analysis, assume that an importing country faces a perfectly elastic export supply schedule. At the given world price, $40 per unit, the country produces 20 units of a commodity and consumes 40 units, resulting in the import of 20 units (Figure 7.6). If the importing country imposes a specific tariff of $5 per unit, the domestic price of the commodity increases from $40 to $45. This increases the country's domestic production from 20 units to 25 units and decreases domestic consumption from 40 units to 35 units. The country's imports decline from 20 units to 10 units.

When the price increases from $40 to $45, consumer surplus declines by area A + B + C + D, and producer surplus increases by area A. The total tariff revenue collected by the importing country is represented by area C, which is equal to area E. The net welfare loss from the tariff, therefore, is equal to the sum of triangular areas B and D. The sum of these triangular areas is also equal to triangular area F in Figure 7.6(b). Area B represents a loss in social utility due to an inefficient increase in production, while area D represents a loss in social utilities due to a decrease in consumption.

In addition to this net welfare loss, the tariff shifts income from consumers to the government and the producers of the protected commodity. The tariff may stimulate growth in the protected industry at a cost paid by consumers. For example, the US steel industry clearly benefits from the high price of steel resulting from a tariff on imported steel. On the other hand, American consumers must pay higher prices on products using steel, such as automobiles and machinery, implying that consumers are worse off under the government protection of the steel industry. A tariff increases

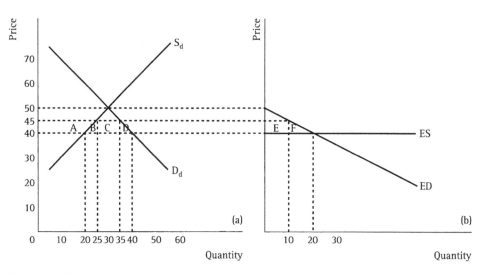

Figure 7.6 The welfare effects of a tariff: a small-country case.

the degree of market power for the protected industry in the country, thereby lowering productive efficiency and penalizing consumers.

■ 7.5 OPTIMAL TARIFF POLICY

Before imposing an import tariff, the world equilibrium is obtained at point E where ES intersects ED in Figure 7.7(b). The quantity traded is 20 units at the world equilibrium price of $40 per unit. If an importing country imposes an import tariff of $15 per unit on a commodity, it is represented by vertical distance ab between ED and ES. The price of the commodity in the importing country rises from $40 per unit to $45 per unit, and the price in the exporting country declines from $40 per unit to $30 per unit, as shown in Figure 7.7. In this case, the total tax revenue collected by the government is equal to area C + E (or C' + E'). This tax revenue is divided into two parts: area C (or C') paid by consumers in the importing country and area E (or E') paid by producers in the exporting country.

The net loss in social welfare in the importing country is area B + D, which is also equal to area G, as discussed in the previous section, and the net loss in the exporting country is area H. The importing country may impose tariffs in such a way that the tax revenue collected from the exporting country, area E (or E'), is large enough to offset its welfare loss (area G). The net gain from an import tariff is the difference between area E (or E') and G. The optimum tariff is the tariff rate that maximizes the net gain from the tariff (E – G). If ES becomes more elastic than ED, the tariff should be reduced to minimize the net loss in social welfare. If ES is horizontal to the x-axis, the tax revenue collected from the exporting country will be zero. Thus,

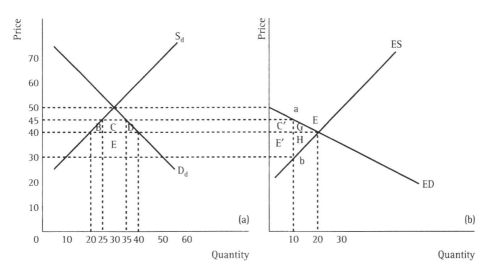

Figure 7.7 The optimal tariff.

the optimal tariff of zero would minimize the net loss in social utility. This implies that the optimal tariff is zero for a small country facing perfectly elastic ES. On the other hand, if ES is sufficiently inelastic compared to ED, the tax revenue collected from producers in the exporting country is larger than the net loss in social welfare. Thus, the government would choose to impose tariffs on the commodity.

If there is a welfare enhancing tariff, the government must decide which tariff rate will provide the greatest gain to society. The tariff rate that accomplishes this is known as the optimal tariff rate.

For luxury goods, ED is generally more elastic than ES, and an importing country's tax revenue collected from producers in the exporting country is larger than the net loss in social welfare given to consumers in the importing country. Thus, the importing country could increase tariffs on luxury goods. On the other hand, the importing country should minimize tariffs on necessities to minimize its welfare loss. In this case, ED is typically less elastic than ES, and the net welfare loss for consumers in the country is larger than the tax revenue collected from producers in the exporting country.

■ 7.6 THE EFFECTIVE RATE OF PROTECTION

Products consumed within a country are either imported, produced completely within the country, or partially produced within the country. Given these differences in the composition of goods, a tariff will have varying effects on the domestic industry and consumers.

Nominal tariffs are applied to the total value of an imported product to protect only the portion of similar products produced at home. When some parts of a domestically produced product are imported from other countries, the total value of the product differs from the value of the product produced at home. Assume that the US television manufacturers import picture tubes and other parts from Japan to produce television sets in the USA. In this case, the value of the television differs from the value added through the domestic production process. The effective rate of protection is the nominal tariff rate applied to the portion of the product produced at home. Hence, the effective tariff rate differs from the nominal tariff rate if the total value of a product differs from the value of the product added through the domestic production process.

Assume that a country levies a 20% import tariff on imported television sets to protect its domestic television industry. Assume further that the imported materials constitute 40% of the final value of the television sets produced at home. This implies that the value of television sets added through the domestic production process is 60%. If television sets sell for $1,000 per unit, the manufacturers spend $400 for imported inputs, and home production adds $600 to the value. The nominal tariff of 20% imposed on imported television sets yields $200 per unit. Since this tariff protects the portion of the value of televisions sets produced at home ($600), the effective rate of protection is 33.3% ($200/$600). This implies that the 20% tariff enables domestic producers of the final product to increase their value added by 40% relative to that of a free trade situation with no tariffs.

If the portion of television value added at home is 80% of the value of televisions ($800 per unit), the effective protection rate is 25% ($200/$800). On the other hand, if the portion of the value added at home is 40% of the value of televisions ($400 per unit), the effective tariff rate is 50% ($200/$400). From this example, effective tariff rates clearly depend upon the portion of the product's value domestically produced and the nominal tariff rates on both inputs and outputs.

The effective rates of protection can be calculated from nominal tariff rates on both inputs and outputs as follows:

$$r_h = \frac{t_h - \alpha_{ih} t_i}{1 - \alpha_{ih}} \tag{7.2}$$

where r_h is the effective tariff rate for commodity h, t_h is the nominal tariff rate for commodity h, t_i is the nominal tariff rate for imported input i, and α_{ih} is the share of imported input i to the total value of commodity h (for the derivation of equation (7.2), see Appendix 7.1).

Continuing with the above example, the effective rate of tariffs can be calculated as

$$r_h = \frac{0.2 - 0.4(0)}{1 - 0.4} = 0.33 \tag{7.3}$$

This indicates that the effective rate of protection is 33% when the nominal tariff is 20% of the value of the products.

When import tariffs on the inputs are 10% of the value of the inputs in the above example, the effective rate of protection is calculated as

$$r_h = \frac{0.2 - 0.4(0.1)}{1 - 0.4} = 0.27$$

The effective rate of protection is 27% when import tariffs are 20% for the final products and 10% for the inputs.

■ 7.7 GENERAL EQUILIBRIUM ANALYSIS OF TARIFFS

We have used a partial equilibrium approach to analyze tariff effects on trade. However, this partial equilibrium approach focuses on a single market and ignores possible effects on the rest of the economy. In this section, a general equilibrium approach of tariff effects on the entire economy is presented. Assume that a country produces two goods, wheat and textiles. Assume further that the country is too small to influence the world prices of these goods. The country's PPF is concave to the origin in Figure 7.8.

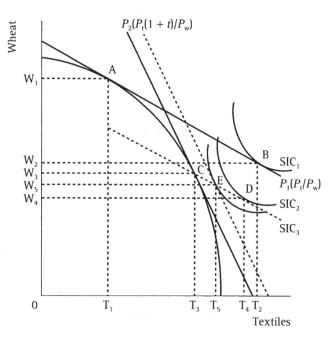

Figure 7.8 The general equilibrium analysis of a tariff in a small country.

Assuming that the equilibrium price of wheat is P_w per bushel and that of textiles is P_t per yard, in an open economy, the terms of trade is P_t/P_w, which is the slope of the country's income (price) line. As demonstrated in Chapter 3, with the given income line, the country's production equilibrium point is obtained at point A where the income line is tangent to the country's PPF curve. The country produces OW_1 units of wheat and OT_1 units of textiles. The country consumes OW_2 units of wheat and OT_2 units of textiles, as indicated by point B, where the income line is tangent to the highest attainable social utility curve of the country (SIC_1). The country exports W_1W_2 units of wheat and imports T_1T_2 units of textiles in a free trade condition.

Now suppose that the country levies an *ad valorem* tariff ($t\%$ of the value of a good) on the imports of textiles. The price of imported textiles rises from P_t to $P_t(1 + t)$, and the slope of the new income line (P_2) is $P_t(1 + t)/P_w$, which is steeper than P_1, indicating that after imposing a tariff of $t\%$, a unit of wheat now exchanges for fewer units of textiles than before. The country produces more units of textiles because of the higher price of textiles, while decreasing its wheat production. With the new income line (P_2), the country produces OW_3 units of wheat and OT_3 units of textiles, as indicated by point C.

Note that the country still faces the international price line (P_1), while consumers face a new domestic price line (P_2). This means that the country trades along the international price line (P_1) at point D, where the price line is tangent to the highest attainable SIC. A shift of SIC from SIC_1 to SIC_2 demonstrates a welfare loss. This loss is due to a reduction in production efficiency as the production point moves from A to C. This loss is equal to triangular area B in Figure 7.6(a). However, point D is not the final equilibrium consumption point, because consumers face a post-tariff price line (P_2). The final equilibrium consumption occurs along P_1 to a point at which the post-tariff price line is tangent to the highest attainable SIC, indicated by point E. A shift of SIC from SIC_2 to SIC_3 represents a welfare loss due to inefficiency in consumption. This welfare loss is equal to triangular area D in Figure 7.6(a). Thus, the total welfare loss due to the *ad valorem* tariff is a shift of SIC from SIC_1 to SIC_3.

The general equilibrium analysis of a tariff in a small country shows two points. First, it demonstrates that trade volume of the two goods decreases with a tariff as compared to free trade. Under a tariff, the country exports W_3W_5 units of wheat and imports T_3T_5 units of textiles, which are much less than those under free trade. Second, social welfare decreases with a tariff as compared to free trade. The social indifference curve shifts inward from SIC_1 with free trade to SIC_3 with a tariff, indicating a reduction in social welfare.

SUMMARY

1 While free trade maximizes world welfare, most countries impose some trade restrictions that benefit protected industries in the country. The most commonly used trade restriction is the tariff. The *ad valorem* tariff is expressed as a percentage of the value of the commodity traded, while the specific tariff is a fixed sum per unit of commodity traded. The compound tariff is a combination of the *ad valorem* and specific duties.

2 If the importing country is large enough to influence the world market, tariffs imposed by an importing country raise the consumer price of the imported commodity in importing countries and decrease the price of the commodity in exporting countries. The incidence of tariffs is concerned with the increase in the price in an importing country relative to the decrease in the export price resulting from the tariffs. The incidence of tariffs depends upon the magnitude of import demand and export supply elasticities. If import demand is more elastic than export supply, more than 50% of the tariffs are borne by producers in the exporting country. Conversely, if export supply is more elastic than import demand, consumers in the importing country bear more than 50% of the tariffs. If import demand and export supply have the same elasticities in the absolute value, the tariffs are shared evenly between consumers in the importing country and producers in the exporting country.

3 If a small country imposes an import tariff, the domestic price of the imported commodity rises by the full amount of the tariff and the world price of the commodity remains the same.

4 The effective rate of protection is defined as the appropriate measure of protection actually provided to domestic producers by imposing nominal tariffs. Thus, as long as the value of a commodity added through the domestic production process differs from the total value of the commodity, the effective rate of protection differs from the nominal tariffs. The two rates are equal when the nominal rate on imported inputs equals the nominal rate on the final commodity or if there are no imported inputs.

5 The nominal tariff is said to be welfare enhancing if the tax revenue collected from its trading partner is large enough to offset its own welfare loss. If the export supply schedule is horizontal to the x-axis, the tax revenue collected from the exporting country is zero

and the welfare loss is large. Thus, the tariff should be minimized to reduce the net loss in social welfare. On the other hand, if export supply is inelastic compared to import demand, the tax revenue collected from producers in the exporting country is larger than the net loss in social welfare. Thus, the government could impose tariffs on the commodity.

6   An optimal tariff is a tariff that provides the greatest welfare gain to society.
7   General equilibrium analysis shows tariff effects on the entire economy, compared to tariff effects on a single sector of the economy in a partial equilibrium analysis. The general equilibrium analysis of tariffs shows reductions in social welfare and trade volume relative to the free trade situation.

KEY CONCEPTS

Ad valorem tariff – A duty measured as a fixed percentage of the price of a good.

C.I.F. – The price of a good at the port in the importing country, including transportation costs, insurance, and loading and unloading charges.

Compound tariff – A duty measured as a combination of *ad valorem* and specific tariffs.

Consumer surplus – The difference between what consumers would have been willing to pay and the market price that they actually pay for a good.

Customs area – A geographical region within which commodities can move freely without restrictions.

Export tariff – A tax or duty imposed on exported goods.

F.O.B. – The price of a good at the port in the exporting country, including loading charges.

Import tariff – A tax or duty imposed on imported goods.

Incidence of tariff – A portion of the duty paid by consumers in the importing country and that by producers in the exporting country.

Nominal tariff – A duty applied to the total value of an imported product.

Producer surplus – The difference between the price that the sellers would have been willing to receive and the market price that they actually receive for a good.

Rate of effective protection – The nominal tariff rate applied to the portion of the product produced at home.

Specific tariff – A duty measured as a fixed amount of money per physical unit of the good.

QUESTIONS AND TASKS FOR REVIEW

1   Explain an *ad valorem*, a specific, and a compound tariff. What are advantages and disadvantages of these tariffs when applied to imported goods?
2   What are the primary purposes of import and export tariffs?
3   Why do most nations exercise protectionism even though they fully understand the benefits of free trade?
4   What is the incidence of a tariff for large and small countries?
5   Using a partial equilibrium analysis, explain changes in social welfare in an importing country when the country imposes an import tariff on an imported good.
6   What are the differences between a nominal tariff and an effective protection rate? How is the rate of effective protection measured?
7   What is optimal tariff policy?

SELECTED BIBLIOGRAPHY

Bhagwati, J. and Ramaswami, V. K. 1963: Domestic distortions, tariffs, and the theory of optimum subsidy. *Journal of Political Economy*, LXXI(1), 44–50.

Greenway, D. 1986: *International Trade Policy: From Tariffs to the New Protectionism*. London: Macmillan Education.

Houck, J. P. 1992: *Elements of Agricultural Trade Policies*. Prospect Heights, IL: Waveland Press.

McCalla, A. and Josling, T. 1985: *Agricultural Policies and World Markets*. New York: Macmillan Press.

Salvatore, D. 2001: *International Economics*, 7th edn. New York: John Wiley.

APPENDIX 7.1 DERIVATION OF THE RELATIONSHIP BETWEEN EFFECTIVE AND NOMINAL TARIFF RATES

The value of commodity h added through the domestic production process without any tariff is

$$V_h = p_h(1 - a_{ih}) = p_h - a_{ih}p_h \tag{A7.1-1}$$

where a_{ih} is the share of imported input i to the total value of commodity h in the absence of tariffs.

Assume that the nominal tariff rate on the imported commodity h is t_h and the nominal tariff rate on the imported input is t_i. The value added in commodity h with tariffs on both the inputs and the outputs is

$$\begin{aligned} V'_h &= p_h(1 + t_h) - p_h a_{ih}(1 + t_i) \\ &= p_h[(1 + t_h) - a_{ih}(1 + t_i)] \end{aligned} \tag{A7.1-2}$$

By definition, the effective tariff rate on the final commodity h (r_h) is

$$r_h = \frac{V'_h - V_h}{V_h} \tag{A7.1-3}$$

Substituting equations (A7.1-1) and (A7.1-2) into (A7.1-3) yields

$$r_h = \frac{t_h - a_{ih}t_i}{1 - a_{ih}} \tag{A7.1-4}$$

Nontariff Trade Barriers

■ 8.0 INTRODUCTION

Nontariff barriers (NTBs) can include any of a number of hindrances other than tariffs – such as policies, rules, and procedures – that distort trade. As multilateral, bilateral, and regional trade agreements decrease tariffs throughout the world, NTBs to trade emerge. NTBs are divided into three types. Type I measures are those where the specific intent is to restrict imports and to stimulate exports in a manner that will inevitably cause trade distortion. Type II measures have the primary intent of dealing with economic, social, and political problems, but are occasionally used to restrict imports and stimulate exports. Type III measures are not intended to be instruments of trade protection, but nevertheless inadvertently cause trade distortion. The NTBs that are most frequently used to control agricultural imports are: (1) quantitative restrictions and similar specific limitations (e.g., quotas, voluntary export restraints, and international cartels); (2) nontariff charges and related policies that affect imports (e.g., antidumping duties and countervailing duties); (3) general government policies that restrict trade (e.g., government procurement policy, competition policies, and state trading); (4) customs procedures and administrative practices (e.g., customs valuation procedures and customs clearance procedures); and (5) technical barriers (health and sanitary regulations and quality standards, safety, industrial standards and regulations, and packing and leveling regulations).

NTBs have not been effectively restricted through multilateral, bilateral, and regional trade agreements. As a result, the incidence of the NTBs appears to be increasing. This chapter identifies and examines some of the most widely used NTBs, including quotas, voluntary export restraints, international cartels, dumping and price discrimination, export subsidies, and technical and administrative barriers.

■ 8.1 QUOTAS

Quotas are absolute quantitative restrictions on exports or imports. Quotas have been banned by the General Agreement on Tariffs and Trade (GATT) and its successor, the World Trade Organization (WTO). The organizations have supported the principle that quantitative restrictions, such as quotas, should be converted into tariffs. This has been encouraged because import and export quotas are more trade distorting than tariffs. Since quotas are more effective than tariffs in insulating the domestic market from the world market, they result in inefficient use of resources in production, and inefficient consumption of goods in the protected industry.

Given this, it is important to provide a description of quota systems and their welfare and economic effects. To accomplish this, an overview of import and export quotas, and their impacts on trade and the domestic market, is provided. Special emphasis is placed on the comparison of import quotas and import tariffs. A description of the tariff rate quota (TRQ), a tool used under various trade agreements to increase market access, is also presented.

IMPORT QUOTAS

Import quotas are quantitative trade restrictions used primarily to protect domestic producers and/or to alter the balance of payments. For example, to protect domestic sugar beet and cane growers, the US government limits sugar imports to approximately 1.25 million tons annually. For an import quota to be effective, the limit must be below what would be imported under free market conditions.

Two major types of import quotas used throughout the world are unilateral quotas and bilateral (or multilateral) quotas. All of these impose absolute limits in value or quantity of imports of a commodity during a given period of time.

The unilateral quota is a fixed amount of imports that the importing country determines without prior consultation or negotiation with other countries. Hence, this type of quota often faces complaints and retaliation from trading partners. In terms of operation of the quota system, the unilateral quota may be global or allocated. A global quota restricts total volume without respect to countries of origin, while an allocated quota assigns import limits to specific trading partners. The US government, for instance, allocates its predetermined imports of sugar to about 40 sugar-exporting countries (an allocated quota), while the Korean government limits its predetermined imports of rice without referencing countries of origin (a global quota).

Under the bilateral or multilateral quota system, the importing country negotiates with exporting countries regarding the quantity that the country should import before determining the allotment of the quota by shares. The importing countries that impose quotas could avoid major complaints and retaliation from trading partners through the use of this type of quota system. This quota could be global or allocated.

EXPORT QUOTAS

Exports may be subject to quantitative restrictions by government action. Export control through an export quota is intended to:

- prevent outflows of strategic goods from the exporting country to the hands of an unfriendly power – the United States controls exports of strategic weapons to prevent outflows of the weapons to countries sponsoring terrorism;
- prevent a shortage of a product in the home market; and
- control the supply of a product to achieve price stability.

Like import quotas, export quotas may be unilateral when the quotas are established without prior agreement with other countries, and may be bilateral or multilateral when they are established through negotiations with other countries. They are typically administered by licensing.

ECONOMIC EFFECTS OF QUOTA

With a downward-sloped import demand schedule and an upward-sloped export supply schedule, as shown in Figure 8.1, an import quota raises the price of the restricted commodity in the importing country and lowers the price in the exporting country. In Figure 8.1, ES is the export supply schedule and ED is the import demand schedule. Under the equilibrium condition where ES intersects ED, the equilibrium price is $50 per unit and the equilibrium quantity is 400 units. When an effective import quota of 300 units is imposed, the price of the commodity increases from $50 to $60 in the importing country and declines from $50 to $40 in the exporting country. Since the importer buys at $40 and sells at $60, the quota profit (rent) is represented by the area A + B, in the absence of transportation costs.

The quota profit, equivalent to $6,000 (= $20 × 300 units), is divided into two parts, as shown in Figure 8.1. The first is paid by consumers in the importing country (area A) and the other is paid by producers in the exporting country (area B). If the importer holds an import license, the importer will have monopoly power and will receive the full amount of the quota profit. On the other hand, if the exporter holds an export license, the exporter will have monopoly power and will receive the quota profit.

If an importing country faces perfectly elastic export supply, an import quota raises the price of the commodity in an importing country and has no impact on the price in an exporting country. In Figure 8.2(a), the price of the commodity increases from $50 to $60 when the importing country imposes an import quota of 300 units. The quota profit, which is equal to area A, is paid by consumers in the importing country. On the other hand, if the exporting country faces a perfectly elastic import demand

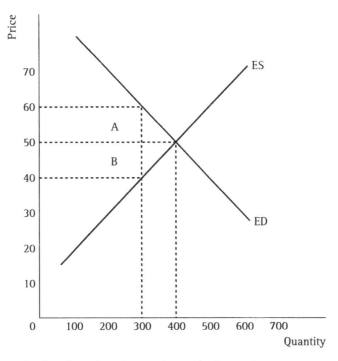

Figure 8.1 Impacts of an import quota on prices: a large-country case.

as shown in Figure 8.2(b), the quota decreases prices in the exporting country, from $50 to $35, indicating that the quota profit (area B) is paid by producers in export-ing countries.

The welfare implications of quotas are similar to those of tariffs. For simplicity, assume that the importing country faces a perfectly elastic export supply schedule. S and D are domestic demand and supply schedules in the importing country in Figure 8.3(a). At a world price of $40, the country produces 200 units of a commodity and consumes 600 units, resulting in the import of 400 units. If the government imposes an import quota of 200 units, the price in the importing country increases from $40 to $50 in Figure 8.3(b). This increase in price results in an increase in domestic pro-duction from 200 units to 300 units and a decrease in domestic consumption from 600 units to 500 units. Trade volume is 200 units, which is equal to the import quota.

As price increases from $40 to $50 in Figure 8.3(a), consumer surplus decreases by area A + B + C + F and producer surplus increases by area A. Area C represents the quota profit, which belongs to importers who hold import licenses. The sum of triangular areas B and F represents the net loss in social welfare resulting from an import quota of 200 units. The sum of these two triangular areas is equal to area E in Figure 8.3(b).

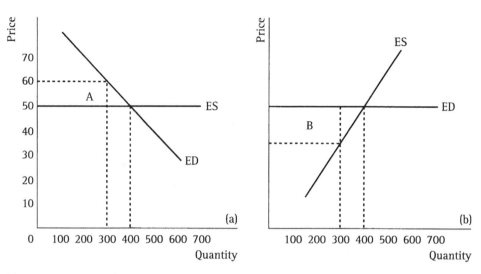

Figure 8.2 Impacts of an import quota on prices in importing and exporting countries: a small-country case.

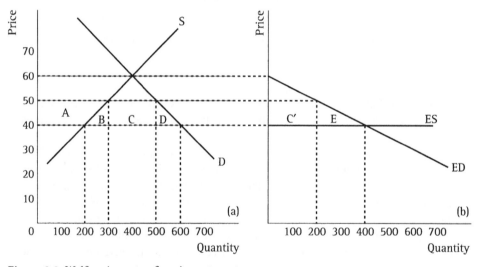

Figure 8.3 Welfare impacts of an import quota.

A COMPARISON OF QUOTAS AND TARIFFS

In contrast to the tariff, the quota profit (or rent) belongs to those who hold an import license. If an importer holds the import license, the profit goes to the importer, who is now able to charge a higher price for each unit of the restricted supply. This is referred to as "monopoly quota profit" (rent), because a monopolist reaps his profit

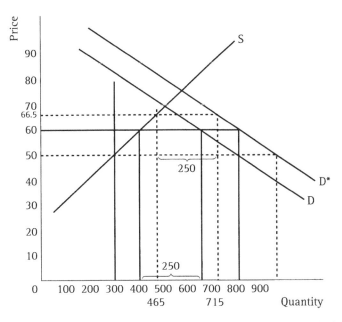

Figure 8.4 A comparison of a quota and a tariff with a shift in demand schedule under perfectly elastic export supply.

in a similar manner by curtailing output and charging a higher price. The domestic price cannot differ from the world price by more than the duty in the tariff system. This is not so in the case of a quota. There is no limit to the price differential between domestic and world prices in a quota system.

Since the quantity of imports is fixed in the quota system, quantity adjustment is not possible. Any changes in domestic demand and supply will cause adjustment in domestic prices. For example, any increase in domestic demand will simply raise the domestic price, leaving imports unchanged. On the other hand, in the tariff system, the domestic price cannot exceed the world price plus the tariff on the commodity. This implies that any change in domestic demand and supply will impact the quantity of imports, leaving prices unchanged. For example, any rise in domestic supply will simply reduce imports, leaving the domestic price unchanged with a tariff.

As shown in Figure 8.4, in the absence of international trade, the quantity produced and consumed in the domestic market is 500 units at the domestic price of $70 per unit. Under free trade, the domestic price is equal to the world price, since the country is too small to influence the world price. At the domestic price of $50 per unit, the country produces 300 units and consumes 800 units, indicating that the country imports 500 units at the price of $50 per unit. If this country levies a specific tariff of $10 per unit, the domestic price increases from $50 per unit to $60 per unit and imports decline from 500 units to 250 units. The same effect would occur on the domestic price and the volume of imports if the government imposed an import quota of 250 units.

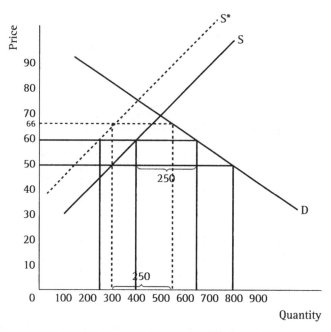

Figure 8.5 A comparison of a quota and a tariff with a shift in supply schedule under perfectly elastic export supply.

Suppose that there is an upward shift in domestic demand to D*. Under a tariff system, the domestic price cannot exceed the world price ($50) plus the tariff ($10 per unit). Thus, the domestic price remains at $60 and the volume of imports rises from 250 units to 400 units since domestic consumption increases from 650 units to 800 units. In other words, the increase in demand is accommodated by an increase in the volume of imports. In the case in which the importing country imposes a quota of 250 units, quantity adjustment is not possible with an increase in domestic demand. The volume of imports is fixed at 250 units. Consequently, the upward shift in demand will produce a price adjustment. The domestic price rises to $66.5, where domestic production plus import (250 units) is equal to increased domestic demand.

Similarly, if domestic production declines due to external factors and the supply curve shifts to S* in Figure 8.5, the country increases imports from 250 units to 400 units at the domestic price of $60 per unit under the tariff system, as shown in Figure 8.5. However, imports remain the same as before (250 units) under the quota system, implying that domestic price adjusts with decreased supply. In the quota system, the domestic price increases from $60 to $66 to reach a new equilibrium with the decreased supply.

Another potential difference between a tariff and a quota is suggested by the theory of effective protection. When a quota is imposed on an imported raw material, it raises the production cost of the final output, of which the material is a part. These effects are the same for tariffs. Import duties on raw materials could be rebated when the final product is exported under the tariff system, but there are no funds to rebate in the case of a quota.

Another difference between tariffs and quotas concerns the case in which the domestic producer of the import substitute is a monopolist. International trade imposes limitations on a firm's monopoly power in a tariff system. In particular, the firm cannot charge more than the world price plus the tariff, because consumers can switch to foreign imports. In the case of an import quota, all an importer needs to do is accommodate a certain fixed amount of imports and, beyond that, the importer is in control of the marketplace. The importer can certainly charge more than with the tariff, since there is less competitive pressure than under the tariff.

BOX 8.1 WHY THE WTO PREFERS TARIFFS TO OTHER TRADE BARRIERS

The original General Agreement on Tariffs and Trade (GATT) laid the foundation for the trade reform that occurred during the second half of the twentieth century. The GATT was designed to increase welfare around the world through the reduction of tariffs and other trade barriers. Despite the consensus that free trade is mutually beneficial, the initial agreement recognized that there are domestic industries in which protection may be needed for various reasons. The agreement set forth the principle that when protection is necessary, it should be provided through the least distorting method possible. Fixed import tariffs meet this criteria. Because of this, the GATT and now the WTO encourages countries to convert nontariff trade barriers into equivalent tariffs and limit the resulting tariffs.

There are several reasons why tariffs are preferred to other trade barriers. The preference to use tariffs rather than other instruments is influenced by the ability of tariffs to provide protection while maintaining many of the essential characteristics of the market. The use of fixed tariffs give traders a clear indication of what duties they must pay to do business. This provides equal access to markets, providing a stable environment for trade to occur. It also allows for clear, although not perfect, transmission of price signals between markets. This is important because it maintains the ability of the market to clearly convey supply and demand information, allowing for an efficient allocation of resources and increased welfare. Finally, tariffs are transparent. The level of protection is easier to determine with tariffs than with other trade barriers. Because of this, negotiations designed to reduce protection levels are less complicated when countries use tariffs to protect domestic industries.

■ 8.2 TARIFF-RATE QUOTAS: A TWO-TIER TARIFF

Another restriction used to protect a domestic industry from foreign competition is the tariff-rate quota. Many governments have imposed tariff-rate quotas as a transition from a quota system to a tariff system. The US government has used tariff-rate quotas on imports such as sugar, milk, and cattle.

The tariff-rate quota is a combination of a tariff and a quota. Imports are subject to different tariff rates associated with the corresponding quotas. For example, under the tariff-rate quota system, the initial imports up to the quota amount would be subject to a tariff of α%, and the volume of imports beyond the quota level would be subject to a higher tariff rate than α%. For example, the US tariff-rate quota quantity for sugar was 1.25 million tons from about 40 countries in 2002. Imports within this limit faced zero duty, but a duty of 16 cents/lb. was imposed on any sugar imports over this limit. This system could be either discriminatory or nondiscriminatory, implying that it might exclude certain exporting countries. Difficulties might occur in administrating this system. The multiple tariffs associated with different import levels require additional accounting records that may be difficult to verify.

Given the import demand and export supply conditions under an assumption that the importing country is large enough to influence the world market, Figure 8.6 shows the quantity traded of 400 units at a world price of $40 per unit. If an importing country imposes *ad valorem* tariffs, the export supply schedule shifts from ES to ES'. The price of the commodity in an importing country increases from $40 to $50, and the price in the exporting country declines from $40 to $30. The total volume traded decreases from 400 units to 300 units. The tariff revenue that the importing country collects is area $A + B + C + D$, equivalent to $6,000 ($20 $\times$ 300 units).

If a quota of 200 units enters duty free and any quantity in excess of this pays an *ad valorem* tax of t%, the supply curve of imports is shown by the bold line in Figure 8.6. The export supply curve is ES when imports are less than 200 units and becomes ES' when imports exceed 200 units. With this export supply curve associated with the TRQ, the market clearing price and quantity are established at $50 and 300 units, respectively. If a quota were used in isolation, the domestic price would rise to $60 and the world price would decrease to $20. Importers would receive a quota profit of area $C + D + E + F$. In the tariff-rate quota case, since the quantity entered duty free is 200 units and an *ad valorem* tax of t% is imposed on imports exceeding 200 units, tariff revenue is area $A + B$. Since the market clearing price is $50 in the tariff-rate quota system, quota profit is equal to area $C + D + F$, which is smaller than the quota profit under the pure quota system.

In the case in which the importing country is too small to influence the world price, the importing country faces a perfectly elastic export supply schedule from the rest of the world, as shown in Figure 8.7. This country imports 600 units at the world price of $15 per unit under free trade. If the country imposes an import quota of 200 units on this commodity, the domestic price increases from $15 per unit to $30

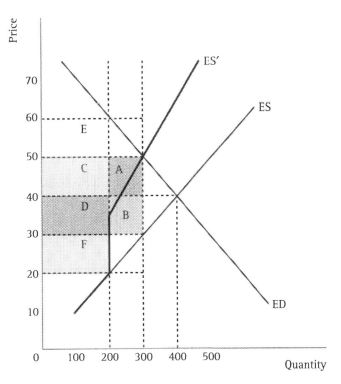

Figure 8.6 The effects of the tariff-rate quota: a large importing country case.

per unit, while the world price remains at $15 per unit. Importers will receive a quota profit represented by area A + C.

Suppose that this country adopts a tariff-rate quota system such that there is no tariff for imports less than 200 units and a tariff of $10 per unit for imports larger than 200 units. The supply schedule of imports with the tariff-rate quota is shown by the bold line in Figure 8.7; the supply schedule is ES for imports less than 200 units and ES ' for imports larger than 200 units. The market equilibrium is obtained at point e', where ES ' intersects the country's import demand schedule (ED'). The country imports 350 units at the price of $25 per unit. Since the first 200 units enter duty free under this tariff-rate quota system, the importer's quota profit is area A, equivalent to $2,000 (200 units × $10), since the importer imports 200 units at the world price of $15 per unit and sells at $25 per unit in the domestic markets. The country also gets a tariff revenue of area B, equivalent to $1,500 (150 units × $10), since the remaining imports (150 units) are charged with a tariff of $10 per unit.

Under the tariff-rate quota system, exporting countries can export within the quota at a tariff lower than the tariff system and still increase exports beyond the quota with a higher tariff. Thus, this system is less restrictive than the quota system. For this reason, this system has been used as a transition from quotas to tariffs.

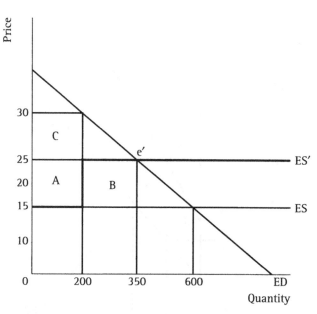

Figure 8.7 The effects of the tariff-rate quota: a small-country case.

■ 8.3 VOLUNTARY EXPORT RESTRAINTS

Voluntary export restraints (VERs) are similar to export quotas in terms of the quantity restrictions in exports. An exporting country voluntarily restricts its exports to a country at a fixed level, which is determined through negotiations. The importing country induces the other country to reduce its exports voluntarily under the threat of higher trade restrictions when the exports threaten the importing country's domestic industry. Since the 1950s, VERs have been used by many developed and developing countries, including the USA and the European Union (EU), to curtail exports of industrial and agricultural products. A classic example involves Japanese export restraints on its automobiles to the USA. From 1977 to 1981, US automobile production decreased about 30% and the share of foreign imports to domestic production increased from 18% to 29%. During the period, about 300,000 workers in the US automobile industry lost their jobs. The US automobile industry lost about $4.0 billion in 1980, partially as a result of automobile imports from Japan. The USA negotiated with Japan and reached an agreement that limited Japanese automobile exports to the USA to 1.68 million units per year from 1981 to 1983, and to 1.85 million units per year from 1984 to 1985.

About 10% of world trade is covered by VERs. Products most affected by VERs include steel and steel products, textiles, agricultural and food products, footwear, and machine tools. Countries/regions protected by VERs are mainly developed countries and trading blocs, including Canada, the EU, and the USA. Exporting countries whose exports have been limited by VERs include Japan, Korea, and Taiwan.

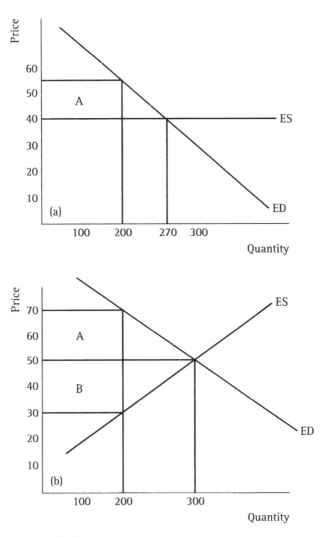

Figure 8.8 The effects of VER: (a) a small-country case; (b) a large-country case.

When an importing country faces a perfectly elastic export supply schedule, the VER raises the price of a commodity in the importing country. In Figure 8.8(a), an importing country imports 270 units at the world price of $40 per unit. If an exporting country voluntarily limits its exports to 200 units, the price of the commodity rises from $40 per unit to $55 per unit in the importing country. On the other hand, VER raises the price of the commodity in the importing country and lowers the price in the exporting country for a large importing country. Under free trade, the importing country imports 300 units at the market clearing price of $50 per unit. If the exporting country restricts its exports at 200 units, the domestic price in the importing country increases from $50 per unit to $70 per unit, and the export price decreases from $50 per unit to $30 per unit. Therefore, the economic effects of VERs are similar to those of quotas. However, there are some differences between VERs and quotas. In the quota system, quota profit (area A in Figure 8.8(a) and area A + B in Figure 8.8(b)) belongs to those who hold the export or import license. Profit goes to exporters in an exporting country with VERs, since VERs are administrated by the exporting countries. Under VERs, an exporter sells 200 units at an export price of $55 per unit in Figure 8.8(a) and 200 units at a price of $70 per unit in Figure 8.8(b). Exporters tend to raise their export prices higher than their domestic prices and capture the quota profits. In addition, import quotas improve the trade balance mainly because world (or export) prices remain unchanged or move in favor of the importing country and the volume of imports is reduced. With a VER, the balance of trade may not be improved, mainly because importing countries pay a higher price for imports that are restricted under VERs.

Another concern regarding the use of VERs is that an exporting country may restrict its exports to the level at which the country can maximize its monopoly profit. While quotas may be global, VERs usually discriminate among countries, which is a violation of the nondiscriminatory rule embodied in the WTO.

VERs are administered by exporters, while quotas are administered by the importing country. This increases the likelihood that VERs could be harmful to the importing country. VERs distort trade flows from exporting countries to importing countries, causing inefficient uses and allocation of resources, and penalize consumers. Under the Uruguay Round Agreement, VERs will be phased out.

■ 8.4 INTERNATIONAL CARTELS AND COMMODITY AGREEMENTS

An international cartel involves collusion among firms located in different countries. The objective is to limit the scope of competition within the market. The Organization of Petroleum Exporting Countries (OPEC) is a cartel that includes most of the oil-producing countries.

Suppose that a group of firms producing a homogenous commodity forms a cartel. The cartel restricts its aggregate exports of the commodity to maximize its profit from sales of that commodity. A cartel management body raises the price of the commodity for its members. However, the cartel's pricing decision may be constrained by the consumer's budget and availability of substitute goods.

A cartel management body can choose any price it likes, but consumers have the final choice to accept or reject goods or services at a stated price. Consumer choice depends upon the nature of the product. If demand for the product is inelastic, consumers accept the commodity regardless of its price. This implies that the cartel could successfully raise the price to maximize its profit. On the other hand, if demand for the product is elastic, consumers may reduce spending on the product as the price of the product increases. For example, since demand for oil is inelastic, OPEC successfully raised the price of oil from $5.00 per barrel to $20 per barrel just after forming the cartel in 1975.

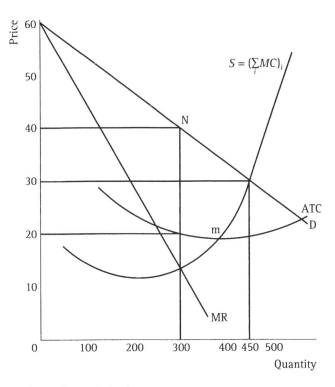

Figure 8.9 A cartel's profit maximization.

If more potential substitutes are available for the product that the cartel produces, demand becomes more elastic and the cartel may be restricted in raising price. However, if potential substitutes are highly limited, the cartel may have more freedom in its ability to maximize profit. For example, the development of oil substitutes may reduce world dependence on oil. OPEC sets the oil price below the price of oil substitutes, to prevent new development of oil substitutes in major oil-consuming countries.

The cartel determines the quantity that it should supply to maximize profits by setting marginal revenue (MR) equal to marginal cost (MC). The cartel's MR is the additional increase in total revenue resulting from one additional unit of output. The cartel's MC is the increase in total cost resulting from one additional unit of output. The cartel's profit is maximized at the output where MR equals MC. As shown in Figure 8.9, the cartel's production level is determined at the point at which its MC equals its MR. To maximize profit, the cartel's total production is 300 units at a market price of $40 per unit at point N, which is higher than the price under a free market condition ($30 per unit). When the cartel's average production is 300 units, its average cost is $20 per unit. Given this, its average profit is equal to $6,000 ($20 × 300 units). Once the cartel decides to produce the quantity of a good that will maximize its profit, the quantity should be allocated amongst its members, using criteria agreed to by the members.

BOX 8.2 THE INTERNATIONAL COFFEE ORGANIZATION

The International Coffee Organization (ICO) is an international, intergovernmental body established in 1963, as a result of decisions made at a United Nations coffee conference. Among other objectives, the ICO was designed to provide a forum for intergovernmental consultation and negotiations on coffee and achieve a reasonable balance between supply and demand. This was designed to ensure that adequate supplies of coffee were available at fair prices to consumers, while profitable markets were available to producers.

The ICO administered a scheme to stabilize coffee prices through a system of export quotas linked to an established price range; a diversification fund to finance projects designed to ration coffee production on a country-by-country basis; and a promotion fund, which financed a number of programs designed to increase coffee consumption.

Before World War II, there was no international action in coffee. Brazil, which accounts for approximately two-thirds of world production, followed a strong price support policy of its own. Other producing countries benefited from this policy and saw no reason to undertake international action. World War II, which prevented producers from shipping coffee to European markets, created the prospect of huge oversupplies. In the face of this situation, an Inter-American Coffee Agreement was signed by the USA and 14 Latin American countries. This agreement entered into force in April 1941. Its main means of supporting coffee prices was a system of export quotas. Initially, the system was effective, but towards the end of the war, quotas were so liberally set that the effect was drastically diminished.

■ 8.5 PRICE DISCRIMINATION AND DUMPING

Another practice that distorts trade flows is dumping or price discrimination. Dumping occurs when a commodity is sold to a foreign buyer at a price lower than the price charged for the identical product on the domestic market. The word "identical" makes it difficult to establish a case of "dumping," because claims must be made for differences in specification, including packaging, in making international price comparisons.

For agricultural products, dumping is a common practice even in industrial countries, including the USA. For example, the agricultural support program, which seeks to maintain prices above their equilibrium market price, results in the accumulation of a large stock of surplus products. The government disposes of the accumulated surplus by selling the products abroad at reduced prices. This is known as government dumping.

Dumping by private companies is divided into three types: sporadic, predatory, and persistent. *Sporadic dumping* is the disposal on foreign markets of an occasional surplus or overstock. The price will go down in foreign markets when a product is sold in the foreign markets below the domestic price. However, the price will return to the equilibrium. Its economic effects are typically negligible.

Predatory dumping occurs when a large, home-based firm sells abroad at a price lower than the domestic price in order to drive out competitors and gain control of the market. After the firm gains control of the market, it intends to reintroduce higher prices and use its newly acquired monopoly power to exploit the market. Potential rivals may then be discouraged from entering the field by the fear of a repeat performance by the monopolist. This may be the most harmful form of dumping.

Persistent dumping occurs when a firm sells a good in a foreign market at a lower price than in the domestic market. Consider a manufacturer who holds a monopoly position in the domestic market and is protected from import competition by transportation costs or government import restrictions. However, the manufacturer faces competition in the foreign market. This implies that the demand elasticity is less elastic in the home market, where the manufacturer has monopoly power, than in foreign markets. The availability of close substitutes in the foreign market makes consumers highly responsive to price change in either direction. To maximize overall net return, the exporter would charge a lower price in foreign markets than in the domestic market. Such dumping is harmful to the foreign producers. This damage may be more than offset by the benefit to its consumers from the lower price.

In Figure 8.10, a firm faces home and foreign markets. It is assumed that demand for the good is inelastic in the home market, while it is elastic in the foreign market. The firm's total marginal revenue is the sum of marginal revenues in the home and foreign markets. To maximize profit, the firm sells 300 units where its MR equals its MC. The firm sells 155 units of the commodity at a price of $73 per unit in the domestic market and sells 145 units at a price of $62 per unit in the foreign market. The difference in price between the home and foreign markets is known as the dumping margin. This dumping is possible if and only if the domestic and foreign markets are separated from each other by means of transportation costs between the two countries and/or trade barriers in the home country. Otherwise, it may be possible for a foreign purchaser to resell the product in the home market. The most common measure to counteract dumping in the importing country is to impose an antidumping duty based on the dumping margin, if the importing country can prove the existence of dumping and show injury to the competing domestic industry. The dumping margin can be calculated by the difference between the normal value of a good and the export price of a similar good, as follows:

$$\text{dumping margin} = \frac{\text{normal value} - \text{export price}}{\text{export price}}$$

The normal value of a good is calculated by the importing country and the export price is the price of a similar good exported by an exporting country.

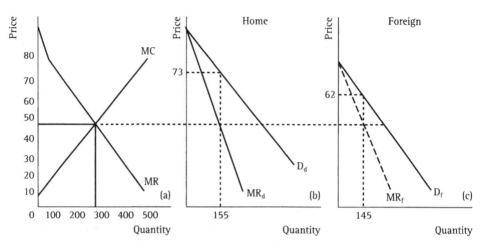

Figure 8.10 Price discrimination between domestic and foreign markets.

■ **8.6 EXPORT SUBSIDIES**

Export subsidies are direct or indirect government payments made to exporters to
stimulate exports of a particular commodity. There are many different types of export
subsidies. These include direct payments, tax relief, subsidized loans, and low-interest
loans. Export subsidies can be regarded as a form of dumping, since exporters can
sell commodities at a price lower than the domestic price under the subsidy program.
The USA used an export enhancement program (EEP) from 1985 to 1995 to promote
US agricultural commodities in the targeted markets. The EU has used its export subsidy program (export refund) to stimulate exports under the Common Agricultural
Policy (CAP). Such export subsidies set prices of agricultural products in the targeted
markets lower than the prices of the same domestic products.

Export subsidies can be described using Figure 8.11. Assume that a small exporting country faces a perfectly elastic import demand. The world equilibrium condition
is obtained at point E, where the country's export supply (ES) intersects the perfectly
elastic import demand (ED). The country exports 200 units at the equilibrium market
price of $50 per unit in Figure 8.11(b). At this price, the country produces 500 units
and consumes 300 units, providing an exportable surplus of 200 units, as shown in
Figure 8.11(a).

If the exporting country gives an export subsidy of $10 per unit, the export
supply schedule shifts outward from ES to ES′, resulting in an increase in exports
from 200 units to 400 units. The increased exports resulting from the export subsidies bid up the price of the export commodity. In Figure 8.11(a), the exporting
country's domestic price would be $60 per unit ($50 + $10), at which the country
produces 600 units and consumes 200 units. The country increases its exports from
200 units to 400 units with the export subsidy of $10 per unit.

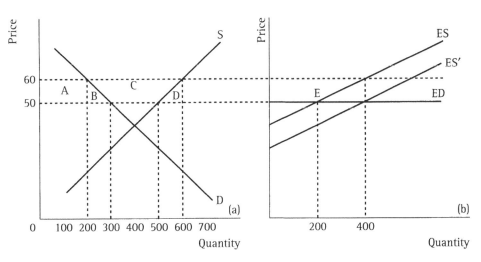

Figure 8.11 The effect of export subsidies: a small-country case.

When the price of the commodity increases from $50 to $60 due to the export subsidy, producer surplus increases by area A + B + C, while consumer surplus decreases by area A + B. The government outlay of the subsidy is area B + C + D. The total losses in social welfare are calculated by subtracting the net increase in producer surplus (area C) from the government outlay (area B + C + D). The net loss in social welfare, therefore, is the sum of the triangular areas B and D. Triangular area B represents the net welfare loss resulting from a reduction in consumption of the commodity caused by the increased price of the commodity. Triangular area D represents the rising domestic cost of producing more units of this commodity.

The effects of export subsidies given by a large country on the world market differ from those for the small country. As discussed above, since a small country faces a perfectly elastic import demand, an increase in its export of a commodity does not change the world price. However, an increase in its exports under the export subsidy affects the world price for the large-country case, since the country faces a downward-sloping import demand. The subsidy given by an exporting country increases domestic production and decreases domestic consumption, resulting in an increase in exports. In Figure 8.12(a), the exporting country exports 200 units at the world price of $50 per unit in free trade. If the exporting country gives an export subsidy of $10 per unit to its exporters, the export supply schedule shifts outward from ES to ES', and increases the country's exports from 200 units to 300 units. The world price decreases from $50 per unit to $45 per unit, and the domestic price increases from $50 per unit to $55 per unit. At the new domestic price, domestic production increases to 550 units and consumption decreases to 250 units, resulting in exports of 300 units.

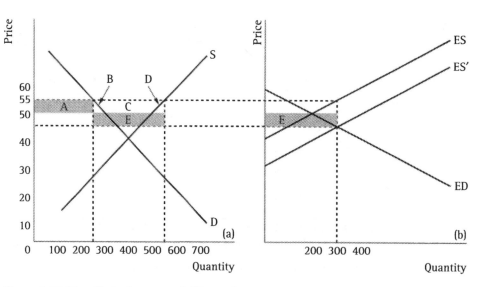

Figure 8.12 The effect of export subsidies: a large-country case.

Since the world price decreases from \$50 to \$45, the export subsidy is beneficial to consumers in the rest of the world. Total benefits given to consumers in importing countries is area E, equal to \$1,500 (\$5 × 300 units) in Figure 8.12(a). The benefits given to producers in the exporting country is area B + C + D, \$1,500 (= \$5 × 300 units). In this case, the benefits from the subsidy were shared between consumers in importing countries and producers in exporting countries evenly. If the price elasticity of import demand (e_m) is more elastic than that of export supply (e_x), the price increase in the exporting country is larger than the price decrease in the world market, indicating that the benefits from the subsidy given to producers in the exporting country are larger than those given to consumers in the world market. Conversely, if the price elasticity of import demand (e_m) is less elastic than the price elasticity of export supply (e_x), consumers in importing countries receive more benefits from the export subsidy than producers in the export country. This implies that an export subsidy program provides benefits not only to producers in the exporting country, but also to consumers in importing countries. However, this export subsidy is harmful to producers in other exporting countries, because of the decreased world price resulting from overproduction.

The total government outlay for this subsidy program is area B + C + D + E. As the price increases from \$50 to \$55, the decrease in consumer surplus is area A + B and the increase in producer surplus is area A + B + C, resulting in a net increase in producer surplus equal to area C. Net losses in social welfare in the exporting country are equal to area B + D (total government outlays (B + C + D) minus the gain in producer surplus (C)). Therefore, the total loss in social welfare is area B + D + E, which is larger than that under a small country's subsidy.

BOX 8.3 THE CURRENT USE OF EXPORT SUBSIDIES IN AGRICULTURE

The Uruguay Round agreement on agriculture imposed disciplines in the areas of export competition, market access, and domestic support. Protection given within these three areas of support was capped with schedules developed for their reduction under the agreement. Within the area of export competition, the use of export subsidies has been targeted as having a destabilizing effect on world markets. While export subsidies are prohibited in other sectors, many developed countries still use export subsidies in agriculture.

With export subsidies, the government pays exporters or producers a fixed or variable amount per unit of exports. These payments increase willingness to sell the product on the world market. This increases the world market supply, thus lowering the world price.

As shown in Table 8.1, the EU paid over 24 billion in US dollars in the form of export subsidies from 1995 to 1998. This accounted for nearly 90% of world export subsidies and imposed large costs on EU taxpayers. Given this expense, why do countries continue export subsidies? The answer is related to the level and type of support provided to the agricultural sector. Developed countries often choose to support their agricultural sectors, resulting in overproduction. To deal with this problem, the government must control imports and encourage exports. Although the cost is very high, providing export subsidies solves a portion of this problem. But the problem is not really solved. It is exported to the rest of the world. Overproduction in the domestic market becomes oversupply on the world market. This lowers world prices and hurts agricultural producers around the world, especially those in developing countries, whose governments cannot afford the levels of agricultural protection given by the EU and the USA.

Table 8.1 WTO member expenditures on export subsidies, 1995–8 (in US dollars)

	Expenditures	Percentage of total
EU	24,369,052,020	89.4
Switzerland	1,402,581,988	5.1
USA	405,976,126	1.5
Norway	341,121,230	1.3
Rest of the world	728,185,008	2.7

Source: Young, L. M., Abbott, P. C. and Leetmaa, S. 2001: Export competition: issues and options in the agricultural negotiations. IATRC Commissioned Paper No. 15, St. Paul, MN

■ 8.7 TECHNICAL, ADMINISTRATIVE, AND OTHER REGULATIONS

International trade is also constrained by numerous technical, administrative, and other regulations imposed by importing countries. These include safety regulations for automobiles and electrical equipment, sanitary regulations for hygienic and agricultural products (e.g., fruits and vegetables), and labeling requirements showing contents. Many of these regulations serve legitimate purposes, but they can also be used to restrict imports from foreign countries.

The Uruguay Round agreements on agricultural commodities addressed sanitary and phytosanitary issues in trading agricultural commodities. The EU bans hormone-treated beef and cattle imports from the USA mainly because the EU claims that the meat is harmful to consumers, even though the beef produced in the USA is widely accepted by both domestic consumers and those in many other countries. The USA

BOX 8.4 NONTARIFF TRADE BARRIERS: THE EU BEEF HORMONE BAN

The EU closed its borders to US beef exports by imposing a ban on beef produced with the help of growth hormones. This action resulted in US producers losing access to a market worth approximately $100 million. The EU's policy ignored the findings of both international and European scientists, which showed growth hormones to be safe when used according to accepted practices.

Countries often use trade restrictions to achieve environmental, health, or other objectives. In achieving these objectives, government policies can be designed to minimize trade distortions. The EU beef hormone ban is an example of a nontariff trade barrier. While it may achieve a social objective, it has the effect of distorting world markets by limiting trade.

After several years of negotiations and litigation, Australia, Canada, New Zealand, and the USA brought their case to the World Trade Organization (WTO). On three occasions, WTO panels ruled that the EU had violated the WTO Sanitary and Phytosanitary (SPS) agreement. In its rulings, the WTO found no credible evidence linking health risks with hormone-treated beef.

Given this ruling, it might be expected that beef exporters now have access to the EU market. Despite losing the case, the EU is not obligated to lift its beef hormone ban. Instead, it has chosen to ignore the ruling to lift the ban. Given that the EU has not complied with the ruling, the USA has the right to impose retaliatory duties against the EU. By imposing these retaliatory duties, the USA may encourage the EU to comply. However, there is no guarantee that this will occur. Despite their success with the WTO panels, US beef producers may continue to be shut out of the EU market.

argues that this import ban by the EU is a type of import restriction, while the EU claims that it is a legitimate sanitary restriction for public health. A similar argument has been made for genetically modified (GM) crops produced in the USA. Some countries and trading blocs, including the EU and Japan, ban imports of GM soybeans from the USA, mainly because they claim that GM soybeans are harmful to consumers, while GM soybeans are widely accepted by consumers in other countries, including the USA.

Other restrictions include laws, known as government procurement policies, that require governments to buy from domestic producers. For example, under the Buy American Act of 1933, the US government gave a price advantage of up to 12% to domestic producers. For defense contracts, the domestic firms had a price advantage of up to 50%. The Tokyo Round of GATT negotiations restricted government procurement policies and gave a fair chance to foreign suppliers.

SUMMARY

1   A quota is a direct quantitative restriction on imports or exports. An import quota has the same effects on consumption and production as an equivalent import tariff.

2   Adjustment to any shift in demand or supply is reflected in the domestic price under an import quota system, while the adjustment occurs in the volume of imports under the tariff system.

3   Another type of quantitative restriction used widely under the Uruguay agreement of the GATT negotiations is the tariff-rate quota (TRQ), which is a combination of the tariff and the quota. There are two different tariff rates. A lower rate is applied to the quantity imported inside the quota and a higher rate to the volume of imports exceeding the quota.

4   Voluntary export restraints (VERs) refer to the case in which an exporting country voluntarily restricts its exports to an importing country through trade negotiations. The economic effects of VERs are similar to those of an equivalent import quota. However, there are several differences between VERs and quotas. In the quota system, quota profit belongs to those who hold import licenses, while the profit goes to exporters with VERs, mainly because VERs are administered by the exporting countries. In addition, import quotas improve the trade balance in the importing country, but this is not necessarily true with VERs. Another difference is that VERs could be discriminatory, while quotas are generally global.

5   International trade is constrained by numerous technical, administrative, and other regulations imposed by importing countries. These include safety regulations, sanitary regulations, and labeling requirements. Although many of these regulations serve legitimate purposes in many countries, they are also used to restrict imports from foreign countries.

6   An international cartel is an organization of suppliers of a commodity located in different countries, that agree to restrict exports of the commodity to maximize their total profits.

7   Export subsidies in an exporting country are a form of direct or indirect government support to exporters in the country, designed to stimulate the country's exports of a particular good. There are many different forms of export subsidies. They include direct payments, tax relief, subsidized loans, and low-interest loans to buyers in foreign countries. Export subsidies raise the domestic price of a commodity. As a result, gains from the subsidy are smaller than the loss to domestic consumers and the cost of the subsidy.

8 If a large exporting country gives an export subsidy, it lowers the world price, implying that the subsidy provides benefits not only to producers in exporting countries but also to consumers in importing countries. However, the subsidy is harmful to producers in other exporting countries.

9 Dumping is the export of a commodity at a price lower than the domestic price. For agricultural goods, dumping is commonly used by both developing and developed countries. Dumping can be persistent, predatory, or sporadic.

10 An importing country can improve an antidumping duty based on dumping margin to protect its import competing industry, if the importing country can prove the existence of dumping and show injury to the industry.

KEY CONCEPTS

Administrative and technical barriers – Import restriction through technical, administrative, and other regulations imposed by importing countries.

Allocated quota – The amount of imports limited to specific trading partners.

Antidumping duty – A duty imposed on an imported good based on the dumping margin.

Bilateral or multilateral import quota – A fixed amount of imports that the importing country determines through consultation or negotiation with other countries.

Dumping – The sale of a good in a foreign market at a price lower than the price charged for the identical good in the domestic market.

Dumping margin – The difference between the normal value of a good and the import price of a similar good.

Export quota – The maximum quantity of exports allowed by the government of an exporting country.

Export subsidies – Direct or indirect government payments to exporters to stimulate exports of a particular commodity.

Global quota – The total amount of imports allowed by the importing country without respect to countries of origin.

Import quota – The maximum quantity of imports allowed by the government of the importing country to protect domestic producers and/or to alter the balance of payments.

International cartels – Collusion among firms located in different countries.

Persistent dumping – A firm sells a good in a foreign market at a lower price than in the domestic market to increase its market shares in the foreign market.

Predatory dumping – A large, home-based firm sells a product abroad at a price lower than the domestic price in order to drive out competitors and gain control of the market.

Quota profit (rent) – The additional profit gained by those who hold import licenses.

Sporadic dumping – The disposal of a good in foreign markets due to an occasional surplus or overstock.

Tariff-rate quota – A trade restriction that combines both a tariff and a quota. Imports are subject to different tariff rates associated with the corresponding quotas.

Unilateral import quota – A fixed amount of imports that the importing country determines without prior consultation or negotiation with other countries.

Voluntary export restraints – The maximum amount of exports that an exporting country agrees to through negotiations.

QUESTIONS AND TASKS FOR REVIEW

1   What is an import quota? What types of quotas have been used by most importing countries?
2   How do the welfare effects of an import quota differ from those of an import tariff?
3   What are the differences in the way in which the import quotas and tariffs adjust to changes in domestic demand and supply?
4   What is a tariff-rate quota? Give some examples of TRQs used by the USA.
5   What is meant by voluntary export restraints? How do they differ from an import quota?
6   What are the technical, administrative, and other barriers to trade? How do they restrict trade?
7   What are international cartels? How does the operation of an international cartel restrict trade?
8   What are the constraints faced by an international cartel?
9   What is international dumping? What are the different types of dumping? What conditions are required to make international dumping possible?
10 What does an importing country do to protect its domestic industry from dumping by foreign firms?
11 Why do countries subsidize exports?
12 What is the impact of an export subsidy given by a large country on the world price and on gains and losses in social welfare?

SELECTED BIBLIOGRAPHY

Deardorff, A. V. and Stern, R. M. 1997: *Measurement of Non-tariff Barriers*. Economic Department Working Paper No. 179. Paris: Organization for Economic Cooperation and Development.

Hillman, J. 1991: *Technical Barriers to Agricultural Trade*. Boulder, CO: Westview Press.

Orden, D. and Roberts, D. (eds.) 1997: *Understanding Technical Barriers to Agricultural Trade*. St. Paul, MN: International Agricultural Trade Research Consortium, University of Minnesota.

Roberts, D., Orden, D. and Josling, T. 2004: *Food Regulation and Trade: Toward a Safe and Open Global Food System*. Washington, DC: Institute for International Economics.

——, Unnevehr, L., Caswell, J., Sheldon, I., Wilson, J., Otsuki, T. and Orden, D. 2001: *The Role of Product Attributes in Agricultural Negotiations*. Commissioned Paper No. 17. St. Paul, MN: International Agricultural Trade Research Consortium, University of Minnesota.

World Trade Organization, *Technical Barriers to Trade*, http://www.wto.org/english/tratop_e/tbt_e/tbt_e.htm. Accessed January 2004.

Young, L. M., Abbott, P. C. and Leetmaa, S. 2001: *Export Competition: Issues and Options in the Agricultural Negotiations*. Commissioned Paper No. 15. St. Paul, MN: International Agricultural Trade Research Consortium, University of Minnesota.

Domestic Support Policies and Trade

■ **9.0 INTRODUCTION**

Since the beginning of Uruguay Round negotiations, which will be discussed in the next chapter, trade negotiators have debated the role of domestic support policies in achieving a freer trading environment. Developed countries and trading blocs, such as the European Union, Japan, and the United States, argue that domestic support should be allowed if it does not excessively distort trade or if it is designed to achieve certain societal goals. Other countries argue that all support distorts trade and should be eliminated. Domestic support is usually a direct cost to the government. Border measures, on the other hand, can be imposed with minimal or zero cost to taxpayers. Given this, developed countries are better able to support their agricultural industries when border measures are reduced or eliminated.

The Uruguay Round of the GATT was revolutionary in the sense that domestic support policies were included in a multilateral trade agreement for the first time. However, the question arises as to what type and level of domestic support should be allowed. Because of this, it is important to consider the trade impact of domestic support policies.

Domestic support is defined as a subsidy given to an industry involved in either exports or imports. One objective of such a subsidy is to benefit the industry facing global competition. Policies such as production subsidies are not directly related to exports or imports of a commodity, and are not classified as trade policy schemes. However, the subsidy often increases domestic supply, resulting in increased exports or decreased imports. Countries have used many different types of domestic subsidies for their domestic industry. In terms of operation, they are classified into income- and price-support programs. This chapter reviews specific subsidies (a typical income-support program) and variable subsidies (a typical price-support program) such as deficiency payments and variable levy systems.

■ 9.1 SPECIFIC PRODUCTION SUBSIDY

Countries give various types of domestic subsidies to their producers in the agricultural sector. Exporting countries give subsidies to make their agricultural sector more competitive in the global market and to enhance net farm income. On the other hand, importing countries often provide domestic subsidies to their producers to reduce agricultural imports as much as possible. Agricultural subsidies are divided into specific (limited) and variable subsidies. Subsidies are fixed in a specific subsidy program. However, subsidies vary with world prices in a variable subsidy program.

DOMESTIC SUBSIDIES IN A SMALL EXPORTING COUNTRY

Under the specific subsidy program, a fixed amount of subsidy is given to producers in a protected industry. The price received by farmers is equal to the world price plus the fixed subsidy. Consequently, the price received by farmers changes in the same way as the world price in this program. The producer's price under this program is calculated as

$$P_s = P_w + S \qquad (9.1)$$

where P_s is the price received by farmers, P_w is the world price, and S is a fixed subsidy. This domestic price received by farmers is higher than the world price by the amount of the subsidy (S). However, consumers in the country pay the world price under this program. Assume that the specific subsidy is $1.00 per bushel of wheat and the world price of wheat is $3.50 per bushel. The price received by farmers is equal to $4.50 ($3.50 + $1.00) per bushel of wheat, while the price paid by consumers is $3.50 per bushel of wheat. As the world price increases from $3.50 to $4.00, the price received by farmers increases to $5.00 ($4.00 + $1.00) and consumer price increases to $4.00 per bushel of wheat.

Consider a small exporting country facing perfectly elastic import demand. As shown in Figure 9.1, domestic production is 50 units of the commodity and consumption is 20 units at the world price of $3.00. The country exports 30 units of the commodities at that price. The corresponding export supply schedule is ES. If the government gives a specific subsidy of $1.00 per unit to producers, the price received by producers increases from $3.00 to $4.00, while the price paid by consumers remains unchanged at $3.00. The domestic supply schedule shifts outward from S to S' with the domestic subsidy, and the export supply schedule shifts from ES to ES'. The vertical distance between these two supply schedules represents the specific subsidy of $1.00 per bushel. The production subsidy given to producers in an exporting country increases production from 50 units to 65 units and export from 30 units to 45 units. The production subsidy does not affect consumers in exporting countries, since the domestic price is equal to the world price in this program.

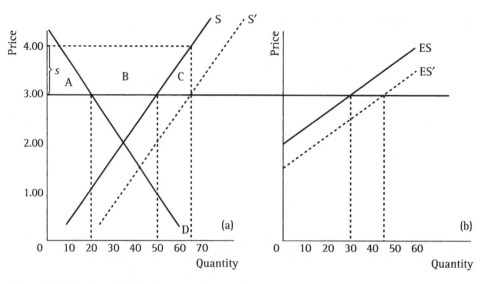

Figure 9.1 Specific subsidies in a small exporting country.

The government outlay is the subsidy per unit (s) times the total quantity of the commodity produced (65 units), which is equal to $65 (area A + B + C). An increase in producer surplus is $57.5 (area A + B). Thus, the net loss in social welfare is equal to $7.5 (triangular area C). Since the price paid by consumers is equal to the world price at $3.00 per unit, there is no change in the consumer surplus.

DOMESTIC SUBSIDIES IN AN IMPORTING COUNTRY

Consider a small importing country facing a perfectly elastic export supply schedule of a commodity, as in Figure 9.2. This country produces 30 units of the commodity and consumes 70 units at the world price of $3.00 per unit. This country imports 40 units of the commodity at this price. With a fixed subsidy equivalent to $1.00 per unit, the domestic price received by farmers increases from $3.00 to $4.00, while the domestic price paid by consumers remains at $3.00 per unit. The domestic subsidy shifts the domestic supply schedule outward from S to S′. Therefore, domestic production increases from 30 units to 45 units, and domestic consumption remains at the same level (70 units). This implies that the production subsidy given to producers in the importing country decreases imports of the commodity by 15 units. As the domestic supply schedule shifts outward from S to S′, the import demand schedule shifts inward from ED to ED′, as shown in Figure 9.2.

Since s is the unit subsidy, the quantity of this product produced in this country is 45 units at the domestic support price of $4.00 per unit. The government outlay is equal to $45 (area A + B). As the producer price increases from $3.00 to $4.00 under

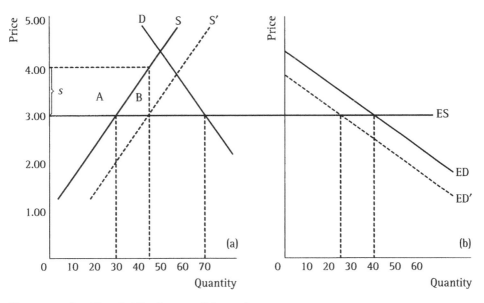

Figure 9.2 Specific subsidies in a small importing country.

the subsidy program, an increase in producer surplus is equal to $37.5 (area A). Thus, the net loss in social welfare is triangular area B, which equals $7.5.

■ 9.2 VARIABLE PRODUCTION SUBSIDY

There are two types of variable production subsidies: the deficiency payment program which the USA has used and the variable levy program which the EU has used. In the variable production subsidy programs, the price paid to producers is fixed at the predetermined level. Thus, the amount of subsidy given to producers changes, depending on the world prices; domestic subsidies increase with decreased world price, and subsidies decrease when the world price increases.

DEFICIENCY PAYMENT PROGRAM

The support price is predetermined at a level higher than the world price. The domestic price is the same as the world price. The deficiency payment is equal to the difference between support and world prices. The government simply pays the differences to producers directly. Therefore, this program does not affect consumers. The unit subsidy given to producers (S) can be calculated as follows:

$$s = P_s - P_w$$

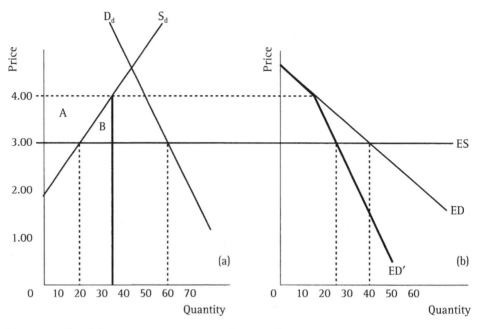

Figure 9.3 The deficiency payment program in a small importing country.

where s is the subsidy received by the farmers, P_s is the predetermined support price, and P_w is the world price. Since P_s is fixed, the domestic subsidy level changes as the world price changes. For example, assume that the support price of wheat is $4.50 per bushel and the world price is $3.00 per bushel. In this case, the subsidy given to producers is $1.50 per bushel of wheat. If the world price increases from $3.00 per bushel to $3.50 per bushel, the subsidy is equal to $1.00 per bushel. On the other hand, if the world price decreases to $2.50 per bushel, the subsidy is equal to $2.00 per bushel.

The objective of variable production subsidies is to guarantee the predetermined support price to producers in the protected industry. Consider a small importing country facing a perfectly elastic export supply schedule in Figure 9.3. The country produces 20 units of a commodity and consumes 60 units of the commodity at the world price of $3.00 per unit under free trade. The country imports 40 units of the commodity. If the government sets the support price at $4.00, domestic production increases from 20 units to 35 units, while domestic consumption remains at the same level (60 units). As a result, the country's imports decrease from 40 units to 25 units. Since domestic producers receive the predetermined price of $4.00 per bushel, the domestic supply remains at 35 units regardless of changes in world prices. The domestic supply schedule is vertical when the world price is less than $4.00 and follows the original supply schedule when world price is above $4.00, as shown in Figure 9.3. The corresponding import demand schedule is ED′. In Figure 9.3, the subsidy program causes imports to decrease from 40 units to 25 units.

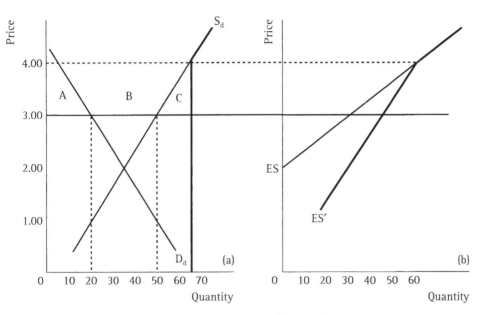

Figure 9.4 The deficiency payment program in a small exporting country.

Since the unit subsidy is the difference between the world price and the support price, and the quantity produced is 35 units, the government outlay is equal to $35 (area A + B). When the domestic producer price increases from $3.00 to $4.00, an increase in producer surplus is equal to $27.5 (area A). Thus, the net loss in social welfare is equal to triangular area B, which is equal to $7.5.

Consider a small exporting country facing perfectly elastic import demand. The country produces 50 units of the commodity and consumes 20 units of the commodity at the world price of $3.00 per unit, as in Figure 9.4. If the exporting country sets a support price at $4.00 per unit, the country increases its domestic production from 50 units to 65 units, while domestic consumption remains at the same level (20 units). The country's exports increase from 30 units to 45 units. Since the government of the exporting country guarantees the support price of $4.00 to producers regardless of the world price, domestic supply remains at 65 units under this production subsidy program, indicating that the domestic supply schedule is vertical for the world price less than $4.00 and follows the original supply schedule for the world price greater than or equal to $4.00. Accordingly, the country's corresponding export supply schedule changes from ES to ES′.

Since the government subsidizes $1.00 per unit to the total domestic production (65 units), the government outlay is equal to $65 (area A + B + C). An increase in producer surplus is equal to $57.5 (area A + B) as the support price increases from $3.00 to $4.00. Thus, the net loss in social welfare is equal to $7.5 (triangular area C).

VARIABLE LEVIES

Governments typically set the domestic price of a commodity at a level higher than the world price under the variable levy system. Variable levies are equal to the pre-determined domestic price minus the world price. The objective of variable levies is to guarantee the predetermined domestic price to producers.

Consider a small importing country facing a perfectly elastic export supply schedule, as in Figure 9.5. At the given world price of $3.00, domestic production is 20 units of a commodity and domestic consumption is 60 units. The country's import is equal to 40 units. If the government sets the domestic price at $4.00 per unit, then domestic production increases from 20 units to 35 units and domestic consumption decreases from 60 units to 50 units, resulting in a decrease in imports from 40 units to 15 units. Since consumers and producers make their consumption and production decisions, respectively, on the basis of the predetermined domestic price, domestic production and consumption remain unchanged regardless of changes in the world price when it is less than $4.00. The corresponding domestic demand and supply schedules are vertical when the world price is less than $4.00 per unit, and follow the original supply and demand schedules when the world price is higher than or equal to $4.00, as shown in Figure 9.5(a). The corresponding import demand schedule is also vertical when the world price is less than $4.00 and follows the original import demand schedule when the world price is greater than or equal to $4.00.

Since the domestic price increases from $3.00 to $4.00 under the variable levy system, this system affects both producers and consumers. This policy causes producer

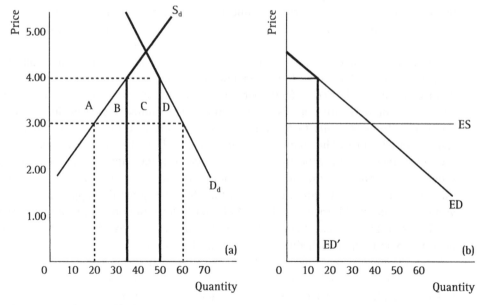

Figure 9.5 The variable levy system in a small importing country.

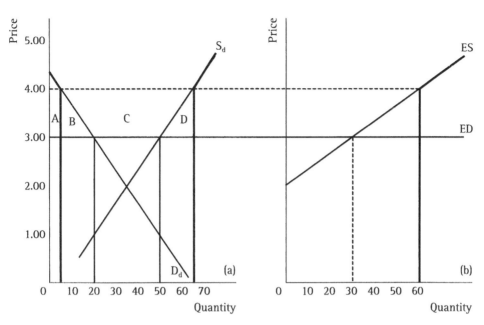

Figure 9.6 The variable levy system in a small exporting country.

surplus to increase, while consumer surplus decreases. The loss in consumer surplus is $55 (area A + B + C + D). The increase in producer surplus is $27.5 (area A). Area C, which is equal to $15, is an income transfer from consumers to producers. The net loss in social welfare is the sum of two triangles, B and D, which is equal to $12.5.

Consider a small exporting country facing perfectly elastic import demand. The country produces 50 units of a commodity and consumes 20 units at the world price of $3.00 per unit, as in Figure 9.6. This country exports 30 units of the commodity. If the government of the exporting country sets the domestic price at $4.00 per unit by imposing a variable levy of $1.00, its domestic production increases from 50 units to 65 units and domestic consumption decreases from 20 units to five units, resulting in an export of 60 units. Since the government sets the domestic price at $4.00 per unit regardless of the world price, the domestic demand and supply schedules are vertical when the world price is less than $4.00 and follow the original demand/supply schedules when the world price is greater than or equal to $4.00, as shown in Figure 9.6(a). The corresponding export supply schedule is also vertical when the world price is less than $4.00 and follows the original export supply schedule when the world price is greater than or equal to $4.00, as shown in Figure 9.6(b).

The loss in consumer surplus is equal to $12.5 (area A + B) as the domestic price increases from $3.00 to $4.00. The increase in producer surplus is equal to $57.5 (area A + B + C). Area C, which is equal to $45, is the net increase in social welfare. The area B + C + D, which is equal to $60, represents the government outlay to subsidize its exports (60 units) since the domestic price is $1.00 higher than the world price. The net decrease in social welfare is the sum of two triangles, B and D, which is equal to $15, under the variable levy program.

A COMPARISON BETWEEN THE DEFICIENCY PAYMENT AND VARIABLE LEVY SYSTEMS

In the deficiency payment system, producers in a country receive subsidies equivalent to the difference between the predetermined support price and the world price, while consumers in the country pay the market price equal to the world price. Consumers in the country are not affected by the subsidy program. The government of the country directly subsidizes producers in the deficiency payment program. However, under the variable levy system, consumers in the country pay the same price as producers in the country receive, implying that consumers in the country directly subsidize producers. There are no the government expenses in the variable levy program in an importing country. However, for an exporting country, the government outlay exists to subsidize its exporters, because the domestic price is higher than the world price and exporters need to have a subsidy to compete in the world market.

Since the domestic price is higher than the world price in the variable levy system, but the same as the world price in the deficiency payment program, the former is more harmful to consumers than the latter and more trade distorting.

■ 9.3 DOMESTIC SUBSIDIES IN A LARGE EXPORTING OR IMPORTING COUNTRY

So far, we have discussed the impacts of various domestic subsidies given to producers in small exporting or importing countries on trade and social welfare. Since the country is assumed to be too small to influence the world market, increases in its production resulting from the domestic subsidies do not affect the world price. The world price remains unchanged when a small country gives domestic subsidies to its producers. However, if a large country gives subsidies to its producers, overproduction of agricultural goods will result. This causes the world price to decrease.

Consider a large exporting country facing a downward-sloping import demand schedule. Domestic subsidies given to producers in the country increase its domestic production. Since the large country is large enough to influence the world price, increased production in the country increases aggregate supply of the commodity in the world market and lowers the world price.

If a large importing country gives domestic subsidies to its producers, domestic production increases. This results in decreases in its imports as well as aggregate demand for the commodity in the world market. The decreased demand lowers the world price of the commodity.

Domestic subsidies in developed and developing countries, including the EU, Japan, South Korea, and the USA, have resulted in over-production of agricultural commodities. This results in the world price decreasing below the production cost.

The elimination of the domestic subsidies would raise the world price of commodities and lower domestic prices in the exporting and importing countries that use the domestic subsidies.

Graphical analyses of various domestic subsidies in a large exporting or importing country are similar to a small-country case as presented in this chapter. In the large-country case, the exporting country faces a downward-sloping import demand schedule and the importing country faces an upward-sloping export supply schedule. A small country faces perfectly elastic import demand or export supply schedules. Students can easily see changes in the world prices resulting from domestic subsidies and impacts on trade flows from Figures 9.1–9.6.

■ 9.4 THE AGGREGATE MEASURE OF SUPPORT

Countries use different types of internal support. The support programs can be summarized through the use of measures such as the producer subsidy equivalent (PSE) and the consumer subsidy equivalent (CSE).

THE PRODUCER SUBSIDY EQUIVALENT (PSE)

The concept of the producer subsidy equivalent (PSE) was developed by the Food and Agriculture Organization (FAO) of the United Nations in the early 1970s. To measure the total value of subsidization provided to producers, the PSE combines direct payments to producers financed by budget expenditures, such as deficiency payments; budget expenditures for programs that provide benefits to farmers, including government-sponsored agricultural research and agricultural inspection services; and the value of revenue transferred to producers from consumers as a result of programs that distort prices, such as border measures or domestic price-support programs. The PSE aggregates support made through various policy tools into a concise measure, allowing agricultural support to be more readily compared across countries.

The PSE represents an aggregate measure of all subsidies given to producers in a country. It is a ratio of aggregate subsidies to total farm revenue. Calculation of the PSE for wheat is illustrated in the following example, where the total revenue from marketing 8 million bushels of wheat at a market price of $4 per bushel is $32 million in Table 9.1. Since wheat producers receive $10 million from their government as a subsidy, their total revenue is $42 million. The PSE for wheat is calculated by dividing the government subsidy by total revenue. In this case, the PSE is 0.24 (10/42), or 24%. The calculated PSE of 24% indicates that without government support, producers would lose the equivalent of 24% of the value of wheat produced with the subsidy in the country. A positive PSE means that taxpayers and/or consumers are paying the subsidy. The subsidies are obtained either directly or indirectly from the taxpayers. A negative PSE means that the producer is giving up that amount, and the government is directly or indirectly taxing the producer.

Table 9.1 An example of producer subsidy equivalent (PSE)

Producer subsidy equivalent is calculated in the following example:
Production	8 million bushels
Market price	$4.00 per bushel
Revenue from market	$32 million (= 8 million × $4.00)
Government subsidy	$10 million
Total farm revenue	$42 million
PSE	0.24 (= 10/42)

Table 9.2 The estimated PSE for all agricultural commodities in selected countries and trading blocs

Country	1986–8 (%)	1999–2001 (%)
Australia	9	5
Canada	34	18
EU	42	36
Japan	62	60
Korea	70	66
Mexico	−1	18
New Zealand	11	1
USA	25	23
OECD average	38	33

Source: OECD, PSE/CSE database, 2002

The PSEs for agricultural commodities in selected OECD countries are shown in Table 9.2. The average PSE for all OECD countries is 33% for the 1999–2001 period, indicating that producers in OECD countries would lose 33% of their agricultural revenue without the government subsidy, assuming that the countries produce the same quantity. PSEs differ among countries; the PSE is highest in Korea, followed by Japan, and lowest in New Zealand. In general, the average PSE for OECD countries decreased from 38% in the 1986–8 period to 33% in 1999–2001 period, indicating a gradual reduction in domestic subsidies in these countries.

THE CONSUMER SUBSIDY EQUIVALENT (CSE)

As opposed to the PSE, the CSE represents an aggregate measure of all subsidies given to consumers in a country. The CSE is defined as the ratio of aggregate government subsidies given to consumers to the total amount paid by consumers. The calculation of the CSE for wheat is illustrated in Table 9.3. Consumers purchase 5 million bushels of wheat at the domestic price of $3 per bushel. Consumers spent a total of $15 million ($3.00 × 5 million), but the world price is $5 per bushel, indicating

Table 9.3 An example of consumer subsidy equivalent (CSE)

Consumer subsidy equivalent is calculated in the following example:
Consumption	5 million bushels
World price	$5.00 per bushel
Domestic price	$3.00 per bushel
Cost to consumer	$15 million (= 5 million × $3.00)
Government subsidy to consumer	$10 million (= 5 million × $2.00)
CSE	0.67 (= 10/15)

Table 9.4 The estimated PSE for all agricultural commodities in selected countries and trading blocs

Country	1986–8 (%)	1999–2001 (%)
Australia	−7	−2
Canada	−22	−14
EU	−39	−31
Japan	−58	−54
Korea	−66	−62
Mexico	18	−14
New Zealand	−9	−4
USA	−7	0
OECD average	−33	−26

Source: OECD, PSE/CSE database, 2002

that consumers receive $10 million from their government ($2.00 × 5 million). The CSE for wheat is 0.67 (10/15), calculated by dividing the government subsidy by consumer spending for wheat. The calculated CSE of 67% indicates that without the government, the price of the commodity would rise by 67%, implying that the government is directly or indirectly subsidizing consumers. A negative CSE means that the government is directly or indirectly penalizing consumers by raising domestic price above the world price. An example of this is a variable levy system, in which the domestic price of a commodity is higher than that of the world price. Thus, consumers are directly penalized under this system.

Average CSEs for selected OECD countries are shown in Table 9.4. The average CSE for the OECD countries is −26% for the 1999–2001 period, indicating that the consumer price of agricultural goods would be 26% lower without government intervention. This implies that governments in OECD countries raised their domestic prices above the world price to protect producers in their countries. For the 1986–8 period, the average CSE for OECD countries was −33%, which was higher than that in the 1999–2001 period (−26%). This implies that agricultural trade has been liberalized over the 1986–2001 period. In the 1999–2001 period, the CSE was highest at −62% in Korea, followed by Japan at −54%. The CSE was zero in the USA in the same period, implying that consumers in the USA are neither subsidized nor penalized.

■ 9.5 DOMESTIC SUPPORT AND THE WTO: THE GREEN, BLUE, AND AMBER BOXES

The Uruguay Round Agreement on Agriculture (URAA) introduced disciplines on domestic support for agriculture. It is widely held that output-promoting domestic subsidies can distort trade patterns even if they do not involve border measures. But for many years countries have not discussed such domestic policies in trade talks or subjected them to trade rules and limitations. The constraints on domestic support measures are still among the most controversial aspects of the WTO negotiations. There are significant disagreements in discussion about the next step in trade policy reform in the WTO negotiations. These disagreements go to the heart of domestic farm policy and the types of instruments that are allowed.

To bring domestic support under the discipline of the GATT, a system of "boxes" corresponding to the color of traffic lights was developed. The red box would include prohibited policies that are not allowed. This box is not currently applied to domestic policies. Yellow or amber box policies are subject to review and reduction over time. Policies in the amber box include market price support and input subsidies. Blue box policies cover payments made with corresponding production limitation programs. An example of blue box policies would be deficiency payments. Finally, the green box includes policies considered to be acceptable and not subject to any limitations. Green box policies are considered to have minimal trade-distorting effects or effects on production. The *peace clause* specifies that domestic support measures that fully conform to the green box provisions are nonactionable and thus exempt from recourse by other countries, including the imposition of countervailing duties. The green box includes both domestic and humanitarian food aid. Developing countries are not required to reduce their protection by the same rate as the DCs. They are also given more time to achieve this reduction.

A number of factors potentially limit the effectiveness of the URAA in reducing support for domestic agriculture. Payments exempt under the green and blue box provisions cover a broad range of support measures. Blue box policies are still widely characterized as market distorting, and many argue that current green box policies also result in distortions. In addition, several countries have been able to meet their URAA obligations by shifting support to exempt (blue and green box) policies without reducing their level of support.

On the other hand, some members feel that the URAA was too restrictive in reducing support for domestic agriculture and neglects "nontrade" concerns. They argue that the survival of their domestic agriculture is necessary for their food security, and to provide a continuing stream of environmental and societal benefits that are joint products of agriculture. These countries typically favor keeping the blue box and adjusting the green box to better accommodate policies that support the multifunctional aspects of agriculture.

BOX 9.1 THE 2002 FARM BILL AND THE WTO

Agricultural policy has traditionally been made by individual countries, with little regard for its impact on other countries. However, agricultural commitments under multilateral trade agreements have become more restrictive. This is increasingly the case for domestic support. Countries are currently free to provide unlimited levels of minimally trade-distorting domestic support within the green box. However, limits have been set on the level of production and trade-distorting agricultural support that can be provided.

The 2002 US Farm Bill was approved in the shadow of the Uruguay Round at a time when the Doha Round of negotiations was just getting started. It is believed by some that this legislation is not consistent with the spirit of the Uruguay Round and that it weakens the US position as an advocate for free trade in the Doha Round.

There are several ways in which the 2002 Farm Act undermines the WTO negotiations.[a] While expenditures are not increased from 1998–2001 levels, they are increased in comparison to the 1996 FAIR Act and extended to include new commodities. Although the 2002 Bill maintains planting flexibility, some of the production and price decoupling achieved in the previous legislation was lost. The 2002 Farm Bill may be legal according to current WTO commitments, but it is not consistent with the free trade position of the US government. Finally, it does not alter the template of rich countries distorting agricultural production and world markets through the provision of high levels of agricultural support.

[a] For a more complete overview of the interdependence of domestic policy and trade agreements, see Orden, D. 2003: *U.S. Agricultural Policy: the 2002 Farm Bill and WTO Doha Round Proposal.* TMD Discussion Paper No. 109. International Food Policy Research Institute, Washington, DC, February; available on the Internet at http://www.ifpri.org/divs/tmd/dp/tmdp109.htm

SUMMARY

1 Domestic subsidies are defined as payments given to producers in an industry in both exporting and importing countries to protect the industry from global competition. Domestic subsidies are divided into specific (limited) and variable subsidies.

2 The subsidy per unit of a commodity is fixed in the specific subsidy program. Under this program, the price of the commodity received by producers is equal to the world price plus the fixed subsidy, while the price paid by consumers remains at the world price. Consequently, the price received by producers changes in the same way as the world price. This program does not affect consumers.

3 In the variable subsidy program, a government sets a support price and pays producers the difference between the support price and the world price. Thus, the subsidy per unit varies as the world price changes. There are two different variable subsidy programs; one is the deficiency program used by the USA and the other is the variable levy system, which the EU has used.

4   In the deficiency payment program, the support price is predetermined at a level higher than the world price and the per unit subsidy is equal to the difference between the support price and the world price. The domestic price remains at a level that is the same as the world price. Thus, the government directly subsidizes producers under the deficiency payment program.

5   Governments set the domestic price at a level higher than the world price in the variable levy system. A per unit subsidy is equal to the difference between the support price and the world price. Under this program, consumers directly subsidize producers by paying a higher price than the world price for the same commodity.

6   The producer subsidy equivalent (PSE) can be used to measure the degree of subsidies given to producers. The PSE represents an aggregate measure of all subsidies given to producers in a country. It is a ratio of aggregate subsidies to total farm revenue. A PSE of 10% in a country indicates that without government intervention, producers would lose the equivalent of 10% of their revenue from agricultural production.

7   The consumer subsidy equivalent (CSE) is the opposite concept to the PSE and represents an aggregate measure of all subsidies given to consumers in a country. The CSE is a ratio of aggregate government subsidies to the total amount paid by consumers. A CSE of 10% indicates that without government intervention, the price of the commodity would rise 10%, implying that the government is directly or indirectly subsidizing consumers.

KEY CONCEPTS

Consumer subsidy equivalent (CSE) – A ratio of aggregate government subsidies given to consumers to the total amount paid by consumers.

Deficiency payment program – A direct payment to producers, which is equal to the difference between the fixed support price and the world price.

Producer subsidy equivalent (PSE) – A ratio of aggregate subsidies given to producers to total farm revenue.

Specific subsidy – A fixed amount of support given to producers in the protected industry.

Variable levy program – A direct payment to producers, which is equal to the difference between the predetermined domestic price and the world price.

Variable subsidy – An amount of support given to producers, which is the difference between the world prices and the support price fixed by government.

QUESTIONS AND TASKS FOR REVIEW

1   How does the specific subsidy program work in exporting countries? How does it affect the export supply schedule in the exporting country?

2   How does the deficiency payment program work in exporting and importing countries, and how does it affect trade of this particular commodity in these countries?

3   How does the variable levy program work in exporting and importing countries, and how does it affect trade of a particular commodity in the countries?

4   Explain the concepts of producer subsidy equivalent (PSE) and consumer subsidy equivalent (CSE).

5   Explain the concepts of green, blue, and amber boxes used in the Uruguay Round of GATT and the WTO negotiations.

SELECTED BIBLIOGRAPHY

Food and Agriculture Organization (FAO) 1975: *Agricultural Protection and Stabilization Policies: A Framework for Measurement in the Context of Agricultural Adjustment*, C 75/LIM/2.

Kennedy, P., Brink, L., Dyck, J. and MacLaren, D. 2001: *Domestic Support: Issues and Options in the Agricultural Negotiations*. Commissioned Paper No. 16. St. Paul, MN: International Agricultural Trade Research Consortium, University of Minnesota.

Orden, D. 2003: *U.S. Agricultural Policy: The 2002 Farm Bill and WTO Doha Round Proposal*, TWD Discussion Paper No. 109, International Food Policy Institute, Washington, DC, February.

——, Paarlberg, R. and Roe, T. 1999: *Policy Reform in American Agriculture: Analysis and Prognosis*. Chicago: The University of Chicago Press.

Sumner, D. A. 1995: *Agricultural Trade Policy: Letting Markets Work*. AEI Studies in Agricultural Policy. Washington, DC: The AEI Press.

CHAPTER 10

Multilateral Trade Negotiations and US Trade Policy

Following World War II, an effort was made to establish an open trading system among nations. The creation of the General Agreement on Tariffs and Trade (GATT) was successful in reducing barriers to trade for sectors such as manufacturing. However, agriculture was an exception.

For a variety of reasons, many developed countries have protected their agricultural sectors through trade barriers and domestic support. At the other extreme, developing countries often tax agricultural exports as a means of collecting revenue and lowering domestic food prices for consumers. In both cases, countries have been reluctant to bring agriculture under the discipline of GATT.

Agriculture was one of the last sectors to be fully integrated within the GATT, and now the WTO. Progress was made toward achieving multilateral trade liberalization in agriculture with the signing of the Uruguay Round agricultural agreement. However, much of the original protection remains.

This chapter reviews the history of multilateral trade negotiations from the Havana Charter through the various rounds of the GATT negotiations and the formation of the WTO in the Uruguay Round. This chapter also provides an overview of the role of the WTO in agricultural trade and potential obstacles to agricultural trade liberalization through multilateral negotiations. US trade policy has developed within the framework established by GATT and WTO. The USA has also participated in multilateral trade negotiations under GATT and WTO, with guidelines provided by the US trade policies. The chapter also explores the evolution of US trade policies to promote free trade and protect domestic industries from unfair trade under multilateral trade negotiations.

■ 10.1 THE GENERAL AGREEMENT ON TARIFFS AND TRADE (GATT)

The trade policies of the United States and other countries were developed mainly within the framework established by the General Agreement on Tariffs and Trade (GATT). GATT is a treaty that was signed in 1947, with 23 countries committed to binding or lowering 45,000 tariff rates on the basis of principles and rules developed under the agreement. GATT became effective in 1948. Since then, the membership of GATT has grown to 123 countries, accounting for about 90% of world trade.

The objective of GATT is to reduce trade barriers and promote fair trade among member countries. GATT is based on the following general principles: (1) concessions to one country must be given to all other GATT member countries (the rule of nondiscrimination in trade relations); (2) imported goods from the member countries should be treated no less favorably than domestically produced goods (the rule of the most favored nation, or MFN); (3) protection for domestic industry should be assured through the use of tariffs, rather than nontariff barriers (prohibitions of quantitative restrictions); and (4) special provisions should be provided to promote trade with developing countries.

GATT focuses on the reduction of tariff rates for individual goods through multilateral trade negotiations and allows most favored nation (MFN) treatment with respect to imports and exports with all member countries. However, GATT includes an escape clause by which members can withdraw or modify a tariff concession if there is an increase in imports resulting from the tariff concession, causing serious injury to domestic producers of the competing industry. Under GATT, members are encouraged to use tariffs rather than quotas to protect their domestic industry because quotas are more trade distorting than tariffs, as discussed in Chapter 8. However, there are exceptions to GATT's prohibition of quotas. GATT allows member nations to use quotas to safeguard their trade balance from a surge in imports in order to promote economic development and the operation of domestic agricultural programs. Voluntary export restraints, a type of import quota, also fall outside the quota resolution of GATT, mainly because they are "voluntary." GATT also allows for the formation of regional free trade areas such as the European Union (EU) and the North American Free Trade Area (NAFTA). Once countries form a regional free trade area, there are no trade restrictions among member countries, but trade restrictions remain for nonmember countries. The trade restrictions imposed on goods from nonmember countries are subject to the disciplines of the GATT.

THE PRIMARY ACTIVITIES OF GATT

The main activities of GATT fall into four areas: (1) reductions in tariffs and nontariff barriers, (2) elimination of quantitative restrictions, (3) settlement of trade disputes, and (4) trade policy review. Each of these activities is intended to achieve the GATT objective of reducing protection and promoting free trade.

Table 10.1 Multilateral trade negotiations under GATT

Date	Name	Number of participating countries	Percentage cut in tariffs
1947	Geneva Round	23	21
1949	Annecy Round	13	2
1950–1	Torquay Round	38	3
1955–6	Geneva Round	26	4
1961–2	Dillon Round	26	2
1963–7	Kennedy Round	62	35
1973–7	Tokyo Round	102	34
1986–93	Uruguay Round	123	34

Source: World Trade Organization Newsletter, Geneva, May 1998

There have been eight multilateral trade negotiations under GATT since its inception. These were the Geneva Round (1947), the Annecy Round (1949), the Torquay Round (1950), the Geneva Round (1955), the Dillion Round (1961), the Kennedy Round (1963), the Tokyo Round (1973), and the Uruguay Round (1986), as shown in Table 10.1. The first negotiations in Geneva resulted in an agreement to reduce 45,000 different tariffs. The following four rounds of negotiation focused on further concessions on tariffs for industrial goods. In the Kennedy and Tokyo Rounds, negotiations included antidumping and nontariff measures. The GATT does not allow any quantitative trade restrictions and it supports the principle that existing quantitative restrictions, such as quotas, should be converted into tariffs. As a result of the focus on the industrial sector, in the first seven rounds of the GATT negotiations, trade barriers for agricultural goods have been much higher than those for industrial goods.

Another aspect of GATT is its role in the settlement of trade disputes between countries. Historically, trade disputes were not resolved in a fair or timely manner because no third party was involved in the process. GATT improved the dispute resolution process by providing a conciliation panel to which a country could express its grievances. However, the GATT dispute process did not have the authority to enforce the panel's recommendations.

THE URUGUAY ROUND OF GATT

The Uruguay Round (UR), the eighth and most ambitious round of GATT negotiations, began in Punta del Este, Uruguay, in September 1986. One hundred and twenty-three countries participated, and the round was completed after 7 years of negotiations. It was scheduled to be completed by December 1990, but it was delayed for 3 years because of disagreements between the USA and the EU, especially France, over the reduction of agricultural subsidies.

In April 1994, most of the 123 participating governments signed the UR agreement in Marrakesh, Morocco. At the meeting, the countries decided to create the World Trade Organization (WTO), an international trade organization responsible for governing the conduct of trade relations among its member countries by administrating the UR Agreement and promoting further liberalization of global trade. The Uruguay Round Agreement was ratified by the US Senate in December 1994. The agreement became effective on July 1, 1995.

In the Uruguay Round, tariffs on industrial goods were to be reduced from an average of 4.7% to 3% and the share of goods with no tariffs was increased from 20–22% to 40–45%. Tariffs were removed altogether on pharmaceuticals, construction equipment, medical equipment, paper products, and steel. Also, many countries agreed to bind or cap a significant portion of their tariffs, giving up the possibility of future rate increases above the bound levels.

Significant progress was also made in decreasing or eliminating nontariff barriers. The elimination of the government procurement policy opened new markets for member countries. The Uruguay Round made an extensive effort to eliminate quotas on agricultural products or replace quotas with tariffs or tariff rate quotas. The safeguards agreement prohibited the use of voluntary export restraints. However, countries were allowed to raise tariffs or other trade restrictions against an import surge that severely harms a domestic industry. The agreement provides 20 years of protection for patents, trademarks, and copyrights.

■ 10.2 ACCOMPLISHMENTS IN AGRICULTURAL TRADE LIBERALIZATION

Special focus was given to liberalization of agricultural commodities under the Uruguay Round. In July 1987, the USA proposed the following agricultural trade reforms to the GATT: (1) a 10-year phase-out of all agricultural subsidies that directly or indirectly distorted trade flows, (2) a phase-out of import barriers over 10 years, and (3) a harmonization of health and sanitary regulations on a basis of internationally agreed standards. The USA also proposed to use an aggregate measurement of support (AMS) and tariffication, respectively, as a procedure to eliminate the various forms of internal subsidies used by countries and the various forms of barriers adopted by importing countries. In December 1988, ministers from participating governments met in Montreal, Canada, for an assessment of progress at the round's halfway point, but the talks ended in a deadlock that was not resolved until officials met again in Geneva the following April. The draft of the "Final Act" was compiled by the GATT director general, Mr. Arthur Dunkel, in December 1991. The draft became the basis for the final agreement. The USA and the EU settled most of their differences on agriculture in the Blair House Accord in November 1992. In addition, the Quad (the USA, the EU, Japan, and Canada) announced significant progress in negotiations on tariffs and market access at the G7 summit in July 1993. After $7^{1}/_{2}$ years of negotiations in the areas of market access, internal support, and export subsidies, the member countries reached an agreement in 1993.

The UR Agreement set rules and commitments for domestic support policies that distort trade flows of agricultural commodities. Developed countries agreed to reduce domestic subsidies in terms of AMS by 20% over 6 years, starting in 1995 (the base level is averages for 1986–8), as shown in Table 10.2. Developing countries agreed to a 13% cut over 10 years. However, a set of policies deemed less trade distorting than others is allocated to the so-called "green box" and is immune to challenge. Other policies are subject to reduction based on the UR Agreement. It was decided that neither the US deficiency payments nor the compensation payments under the reformed Common Agricultural Policy (CAP) of the EU needed to be changed.

The participating countries agreed to convert all existing nontariff barriers into bound duties (tariffication) and not to introduce new nontariff measures. Tariffs were to be reduced by an average of 36% in developed countries for the 6 years starting in 1995 and by 24% over 10 years in developing countries (the base level is an average of the years 1986–8). "Tariffication" imposed policies for a number of countries; Canada agreed to replace import quotas for poultry products with tariffs and the EU agreed to replace its variable levy with tariffs. The USA forwent the use of its Section 22 import quotas.

Under minimum market access, the import quotas were expanded to 5% of domestic consumption over the 6-year period in developed countries and over the 10-year period in developing countries. Japan and Korea agreed to expand market access for rice in compensation for a delay in introducing tariffs.

Countries accepted commitments on reducing spending on export subsidies and the quantity of subsidized exports. Developed countries agreed to cut the value of export subsidies by 36% over the 6 years (the base level is the average for 1986–8), while developing countries were committed to 24% over 10 years (Table 10.2). The quantities of subsidized exports would be cut by 21% over 6 years in developed

Table 10.2 Uruguay Round targets (percentages) for subsidy and protection reductions[a]

	Developed countries; 6 years, 1995–2000	Developing countries; 10 years, 1995–2004
Market access		
Average tariff cut	36	24
Minimum cut per product	15	10
Export subsidies		
Cut in value	36	24
Cut in quantity	21	14
Domestic support		
Total aggregate measure of support cuts	20	13

[a] Note that the base for market access and domestic support is the 1986–8 average. The base for export subsidies is the 1986–90 average.
Source: World Trade Organization

countries and by 14% over 10 years in developing countries. Countries also agreed not to apply export subsidies to commodities not subsidized during the base period.

Many countries have been engaged in a process of reducing government support to agriculture and making such support more closely targeted to needs in advance of the outcome of the Round. Policy reforms in the EU, Canada, Sweden, Australia, New Zealand, and Latin America have been strongly influenced by the UR Agreement. Reforms have occurred to a lesser extent in Japan and Korea. The 1996 FAIR Act was considered part of the process of global trade liberalization under the UR Agreement.

10.3 THE WORLD TRADE ORGANIZATION (WTO)

The WTO was created in January 1995 to succeed GATT. The WTO is responsible for governing the conduct of trade relations among its 148 members. GATT obligations remain at the core of the WTO. However, the WTO Agreement requires its members to adhere not only to GATT rules but also to the broad range of trade pacts that have been negotiated under GATT. The WTO regulates international trade activities among the member nations on the basis of the following principles: (1) nondiscrimination embodied in the most favored nation clause; (2) a general prohibition of export subsidies and import quotas; and (3) a requirement to retain or reduce an average tariff rate. Developing countries and agricultural product trade are exempt from some of these rules.

Unlike GATT, the WTO is a full-fledged international organization, headquartered in Geneva, Switzerland. WTO functions are much broader than those of GATT. In addition to overseeing rules related to the international trade of goods, the organization deals with commercial services, intellectual property rights, and foreign investments. It also deals with trade disputes among the member countries.

Through various councils and committees, the WTO administers many agreements contained under the Uruguay Round Agreement. It oversees the implementation of the tariff cuts and reductions in nontariff measures agreed to in the negotiations. It also examines trade regimes of individual members. Members are also required to provide updated trade measures and statistics, which are maintained by the WTO.

A major function of the WTO was strengthening the GATT's procedure for settling trade disputes. The GATT dispute mechanism suffered from long delays and inadequate enforcement. The dispute-settlement procedure of the WTO addresses each of these weaknesses. The WTO guarantees the formation of a dispute panel once a case is brought before the organization. Once a decision is made by the panel, the accused country can no longer block the panel's final decision.

The WTO initiated its first multilateral negotiations in 1999. However, the negotiations have not been successful, mainly because of conflicting agendas among the member countries.

THE MINISTERIAL MEETING IN SEATTLE

WTO negotiations started in Seattle in 1999. However, the meeting collapsed because of the conflicting agendas of its participants, mainly the USA, the EU, Japan, and the developing countries. In addition, there were demonstrations by antiglobalization activists, including trades unionists, environmentalists, and human rights activists opposed to WTO policies. US labor unions opposed to the organization argue that firms in developing countries must meet labor standards similar to those in the USA if they are to sell their products in this country. Environmentalists insist that globalization has increased pollution in developing countries, where environmental regulations are not as strict as in developed countries. Finally, human rights activists demand that all participating countries respect human rights and treat their citizens with dignity.

In addition, developing countries have actively participated in the negotiation process by demanding a relaxation of antidumping enforcement, the creation of special trade rules for developing countries, and the opening of export markets for labor-intensive goods, such as farm goods, textiles, and clothing. These demands directly conflict with those of developed countries, mainly the USA and the EU. Developing countries oppose the US proposal on protections for labor rights and the environment. Other items in the US proposal include termination of the EU's export subsidies and reductions in tariffs on goods and services. Because of the differences in agendas between developed and developing countries, these WTO meetings ended without reaching a compromise.

THE DOHA MINISTERIAL MEETING AND THE DOHA DEVELOPMENT AGENDA (DDA)

The Doha Ministerial Conference in November 2001 established the agenda for a new round of WTO trade negotiations. This was the second attempt to launch a new round of negotiations since the WTO was created during the Uruguay Round in 1994. The Ministerial Conference in Doha, Qatar, in November 2001, was more successful than the conference in Seattle. The Doha Declaration provides the mandate for negotiations on a range of subjects, including the implementation of present agreements. The entire agenda is referred to as the Doha Development Agenda (DDA).

Subsidiary negotiating bodies were created to handle individual subjects. Twenty subjects were listed in the Doha Declaration; most of these topics involve the process of negotiations, while some include actions for implementation, analysis, and monitoring. In addition to implementation-related issues, the subjects include agriculture; services; market access for nonagricultural products; trade-related aspects of intellectual property rights (TRIPS); the relationship between trade and investment; the interaction between trade and competition policy; transparency in government procurement; trade facilitation; antidumping and subsidies; WTO rules for regional trade agreements; dispute settlement understanding; trade and the environment;

electronic commerce; small economies; trade, debt, and finance; trade and transfer of technology; technical cooperation and capacity building; least-developed countries; and special and differential treatment.

The Doha Declaration set a timetable for these negotiations, with a final deadline of January 1, 2005. Intermediate deadlines were also created. The Doha Agenda outlined a number of tasks to be completed before or at the Fifth Ministerial Conference, which was held in Cancun, Mexico, in September 2003. A few of these tasks were completed, such as agreement on the TRIPS and public health issues, but many of the deadlines were missed. The conference in Cancun was not successful in resolving many of these issues.

CANCUN MINISTERIAL MEETING IN 2003

The WTO Ministerial Conference held in Cancun, Mexico, on September 10–14, was not able to reach a consensus because of differences in interests among the member countries, mainly the USA, the EU, Japan and the Cairns Group, including a number of developing countries. The conference dealt with the modalities phase of negotiations, which was originally scheduled to be completed by March 31, 2003. The purpose of this phase of negotiations is to set targets, including numerical targets, for achieving the objectives set out in the Doha Ministerial Declaration. The three major areas of negotiation are market access, export competition, and domestic support. The modalities would describe how the final trade agreement would be shaped. However, a settlement on the modalities could not be reached by the March 31 deadline. Many hoped that negotiations at the Cancun conference would result in a decision, but an agreement once again proved elusive. Besides differences on agricultural issues, there were also disagreements between the WTO members on so-called "Singapore issues," which contributed to the collapse in negotiations. These issues include investment rules, competition policy, transparency in government procurement, and trade facilitation.

The negotiations failed in March mainly because of large differences between the USA, the EU, Japan, the Cairns Group, and other countries on issues such as tariffs, export subsidies, and domestic support. A month prior to the Cancun conference, the USA and the EU reached a framework agreement for negotiating these agricultural issues. In response, a group of 21 developing countries, including Brazil, China, and India, provided their own framework. Additional frameworks were proposed by other groups of countries. Developing countries claim that the US–EU plan, which calls for reduction of the most trade-distorting domestic support measures but allows less trade-distorting support to be maintained, does not include enough of a reduction in subsidies. The group of developing countries wants the blue box eliminated and a cap or reduction on the green box for developed countries. The USA argues that it is willing to cut subsidies, but that the proposal by the developing countries does not offer enough in return in terms of market access. After hearing the positions of the WTO members, George Yeo Yong-Bon, the facilitator of the agricultural negotiations at Cancun, wrote a draft text in the hopes of bridging the gaps and reaching a consensus. Some progress was made, but a consensus could not be reached, as many members found the draft either too ambitious or not ambitious enough.

■ 10.4 US TRADE POLICY AND TRADE REMEDY LAW UNDER GATT AND WTO

US trade policy has developed within the framework established by GATT and WTO. The USA has also initiated some of the multilateral trade negotiations under US trade policy. This section explores the evolution of US trade policy since 1930 and its interaction with multilateral trade negotiations under GATT and WTO.

THE EVOLUTION OF US COMMERCIAL POLICY

The Smoot–Hawley Act was passed by the US Congress and signed by President Herbert Hoover in 1930, despite formal protests from foreign nations and economists around the world. The purpose of this bill was to stimulate the domestic economy by restricting imports from foreign countries. Under the bill, US average tariffs were raised to 53% on all imports. The legislation provoked retaliation by 25 trading partners of the USA. As a result, US trade with other nations collapsed. The Great Depression and foreign retaliations on US exports resulted in a reversal of US trade policy.

In order to stimulate the US economy from the deep recession, the US Congress passed the Reciprocal Trade Agreements Act in 1934 to reverse US trade policy under the Smoot–Hawley Act. Under this law, the President was given authority to negotiate bilateral tariff reduction with foreign governments. The President, under this Act, could lower tariffs by up to 50% of the existing level without approval from the US Congress. The Act also provided for tariff reduction through the most favored nation (MFN) clause. Under this Act, the USA participated in the four rounds of GATT negotiations.

In 1962, the Trade Expansion Act was passed by the US Congress to replace the Reciprocal Trade Agreement Act, in order to deal with the new situation created by the formation of the Common Market in Europe. The Trade Expansion Act authorized the President to negotiate average tariff reductions of up to 50% of the 1962 level. This Act also provided adjustment assistance to displaced workers and firms injured by tariff reductions. Under the authority of this Act, the USA initiated the Kennedy Round of GATT negotiations, which started in 1963 and was completed in 1967. Negotiations resulted in an average tariff reduction on industrial products by a total of 35% of their 1962 level.

The Trade Expansion Act of 1962 was replaced by the Trade Reform Act in 1974. This Act authorized the President to negotiate an average tariff reduction of up to 60%, remove tariffs of 5% or less, and negotiate reductions in nontariff trade barriers. Under the authority of this Act, the USA participated in the Tokyo Round of

GATT negotiations, which started in 1973 and was completed in 1977. Under the agreement, average tariffs on industrial goods were reduced by 34% over an 8-year period, starting in 1980. The negotiations also reached an agreement on nontariff barriers.

More recently, the US Congress passed the Trade and Competitiveness Act of 1988. This Act authorized the USA to participate in the Uruguay Round of GATT negotiations, which started in 1988 and concluded in 1993. As discussed previously, the Uruguay Round reduced average tariff rates by 34% during the subsequent 6–10 years. This Act includes a variety of provisions that are designed to protect the domestic industries under global competition. The provisions deal with unfair trade under Section 301 and trade relief under Section 201. The provision in Section 301 requires the administration to publicly list countries that trade unfairly with the USA, to negotiate removal of such practices within 3 years, and to take retaliatory action if negotiations fail. Under Section 201, the President can impose duties and other trade restrictions on imports for a limited period of time if imports cause serious injury to the domestic industry.

PROTECTION THROUGH TRADE REMEDY LAWS

In addition to multilateral trade negotiations for free and fair trade practices, the USA has adopted a series of trade remedy laws designed to protect domestic industries from unfair foreign competition. The trade remedy laws that have been commonly used by the US government are the antidumping law, the countervailing duty law, the Agricultural Adjustment Act (Section 22), unfair trade under Section 301 of the Trade Act of 1974, and the escape clause under Section 201 of the Trade Act of 1974.

Antidumping duties

The antidumping duty is imposed to offset unfair trading practices by foreign firms that either sell a product in the USA at prices below the average production cost of the product or sell the product in the USA at prices below prices that the exporters charge for comparable products in their home markets. Dumping generally refers to a form of international price discrimination. These trading practices by foreign firms are harmful for the US industry or firms that produce comparable products. The US industry or firms that are injured as a result of the unfair trading practices can file a petition with the US International Trade Commission (USITC).

Upon receipt of a petition, the USITC conducts an investigation to determine whether the dumping has occurred and if it causes material injury to the domestic industry or firms. If the investigation finds that the foreign exporters are practicing dumping, then the US government imposes an antidumping duty equal to the margin of dumping, which is the difference between the price of a foreign good and the domestic price of a similar good produced in the USA.

Countervailing duties

The purpose of the countervailing duty law is to offset any unfair competitive advantage that foreign manufacturers (or exporters) might enjoy over US domestic producers as a result of subsidies. A foreign firm that receives an export subsidy may be able to sell a commodity at a price lower than the price that domestic firms can charge for the identical commodity that is not subsidized. This foreign producer would have an unfair advantage over the domestic producers. Domestic producers injured as a result of unfair trade practices can file a petition with the US Department of Commerce (DOC).

The DOC conducts a preliminary investigation of a petition filed by a US industry. If the preliminary investigation finds reasonable indication of an export subsidy, the US government imposes a special tariff equal to the estimated subsidy margin on imports of the product. The DOC continues to conduct the investigation to determine whether an export subsidy gives an unfair advantage to foreign firms. If it determines that there was no export subsidy, the special tariffs imposed on the imports are rebated to importers of the product. If the investigation by the DOC finds that foreign firms received an export subsidy, then the USITC determines whether the domestic industry is materially injured or is threatened with material injury. If the USITC finds material injury, a permanent countervailing duty is imposed. The duty should be equal to the size of the subsidy margin calculated by the DOC in its final investigation.

The escape clause (Section 201)

The escape clause provides relief to US firms and workers desiring protection from increased imports stemming from trade negotiations. The President has the authority to withdraw from and modify trade concessions granted to foreign countries, and can impose duties and other trade restrictions on imports of any product for a limited period of time if such imports are proven to cause serious injury to the domestic industry.

An escape clause action is usually initiated by a petition by an industry to the USITC. The USITC investigates and recommends a decision to the President, who determines remedy for the injured industry.

Unfair trade (Section 301)

Section 301 provides the US Trade Representative (USTR) the authority to respond to certain unfair trade practices by foreign nations. If the USTR determines that a foreign act, policy, or practice violates or is inconsistent with a trade agreement, or is unjustifiable and burdens US commerce, the USTR has discretionary authority to enforce the trade agreement or to eliminate the act, policy, or practice. In addition, the USTR imposes tariffs or other import restrictions on foreign products if the USTR determines that a foreign nation is engaged in unfair trading practices.

BOX 10.1 UNFAIR TRADING PRACTICES

The North Dakota Wheat Commission alleged that the Canadian Wheat Board, a state trading enterprise with a near monopoly on Canadian wheat sales, engaged in unfair trade practices in its export sales of wheat to the US market and to certain third-country markets of interest to US exporters. In September 2000, the North Dakota Wheat Commission filed a petition under Section 301 of the *Trade Act* of 1974 requesting an investigation of the wheat marketing practices used by the Canadian Wheat Board. The United States Trade Representative (USTR) undertook a 16-month investigation and requested that the USITC examine the competitive practices of the Canadian Wheat Board in the US market and overseas. At the request of the USTR, the Commission instituted investigation No. 332–429, concerning the conditions of competition between US and Canadian wheat in the USA, and in certain third-country markets. Two types of wheat – hard red spring (HRS) and durum – were included. The USTR requested that the USITC submit its confidential report to the USTR by September 24, 2001, later extended to November 1, 2001.

In February 2002, the USTR released the findings of their investigation. The report indicated that the Canadian Wheat Board had used special monopoly rights and privileges that disadvantaged US farmers and were unfair to trade and had, in effect, been taking sales from US farmers. As a result of these findings, the USTR decided to aggressively attempt to level the playing field for US farmers. The USTR announced that it would examine the possibility of filing US countervailing duty and antidumping petitions with the US Department of Commerce (DOC) and the USITC.

On September 13, 2002, a petition was filed with the USITC and DOC by the North Dakota Wheat Commission; the Durum Growers Trade Action Committee; and the US Durum Growers Association. alleging that industries in the USA were materially injured and threatened with material injury by reason of subsidized and less than fair value (LTFV) imports of durum and HRS wheat from Canada. On October 23, 2002, the International Trade Administration (ITA) and the DOC announced its decision to initiate antidumping (AD) and countervailing duty (CVD) investigations on imports of durum and HRS wheat from Canada.

According to a preliminary determination on November 25, 2002, the USITC determined that there was a reasonable indication that the subject industries in the USA were being materially injured by reason of imports from Canada of durum and HRS wheat. The legal standard for preliminary antidumping and countervailing duty determinations requires the Commission to determine, on the basis of the information available at the time of the preliminary determinations, whether there is a reasonable indication that a domestic industry is materially injured, threatened with material injury, or whether the establishment of an industry is materially retarded, by reason of the allegedly unfairly traded imports.

After an affirmative preliminary determination by the USITC, the DOC imposed a preliminary tariff of 3.94%, as of March 4, 2003, on all varieties of spring wheat from Canada. The DOC also imposed preliminary AD duties of 8.15% on durum wheat and 6.12% on HRS wheat imports from Canada on May 2, 2003.

Section 22 of the Agricultural Adjustment Act of 1933

This Act authorizes the President to impose fees or quotas on imported products that undermine any US agricultural commodity program. This authority was designed to prevent imports from interfering with USDA efforts to stabilize domestic agricultural commodity prices. On the basis of USITC investigations, the President determines whether statutory conditions warranting the imposition of a Section 22 or fee exist.

If the President makes an affirmative determination, either import fees or import quotas are imposed to prevent imports of the product concerned from harming or interfering with the relevant agricultural program. In the UR Agreement on agriculture, Section 22 authority is confined in its applicability to imports from countries that are not WTO members.

BOX 10.2 APPLICATION OF SECTION 22 OF THE AGRICULTURAL ADJUSTMENT ACT

In 1993, the US wheat industry requested the Clinton Administration to take legal action under the provisions of Section 22 of the *Agricultural Adjustment Act (AAA)*, as amended. As directed by the President, the USITC instituted investigation No. 22–54 on November 17, 1993, under Section 22(a) of the Agricultural Adjustment Act of 1933, to determine whether wheat was imported into the USA under such conditions or in such quantities as to render or tend to render ineffective, or materially interfere with, the price support, payment, and production adjustment programs conducted by the USDA for wheat.

The USITC determined, by majority rule, that wheat, wheat flour, and semolina were being imported into the USA under such conditions and in such quantities as to "materially interfere" with the price-support programs conducted by the USDA for wheat. The Commission's report to the President indicates that Canadian exports of wheat interfered with the US price support program and led to a negotiated settlement for the 1994–5 crop year, which is known as the Wheat Peace Agreement. For the 12-month period beginning September 12, 1994, the USA applied the following tariff rate quota on the importation of wheat into the U.S.: NAFTA rate for durum wheat imports ranging from 0 to 300,000 metric tons, a tariff of $23 per ton for imports ranging from 300,000 to 450,000 metric tons, and a tariff of $50 per ton for durum wheat imports larger than 450,000 metric tons; and NAFTA rate for imports of other wheat ranges from 0 to 1,050,000 metric tons, and a tariff of $50 for imports of other wheat larger than 1,050,000 metric tons. In market response to the agreement, the US domestic price of durum wheat rose from $4.67 per bushel in 1994 to $5.75 per bushel in 1995. However, the price fell to $3.95 per bushel in 1997.

SUMMARY

1 The General Agreement on Tariffs and Trade (GATT) was established in 1947, with a membership of 23 countries. The objective of GATT was to reduce trade barriers and promote free trade among member countries. Since the establishment of GATT, tariffs for industrial goods have been reduced through eight rounds of GATT negotiations. The eighth round of GATT negotiations (the Uruguay Round) brought agricultural trade into agreement for the first time.

2 In the Uruguay Round of GATT negotiations, member countries agreed to reduce not only tariffs, but also export subsidies and domestic subsidies for agricultural goods.

3 The World Trade Organization (WTO) was created in 1995 under the Uruguay Round Agreement, to succeed GATT. The WTO functions are much broader than those of GATT. It not only oversees the institutional framework for multilateral trade negotiations to promote freer trade among the member countries, but it also deals with commercial service, intellectual property rights, and foreign investments.

4 US commercial policy has been aimed at promoting free and fair trade for agricultural and industrial goods by enacting several trade laws. These include the Reciprocal Trade Agreement Act of 1934, the Trade Expansion Act of 1962, the Trade Reform Act of 1974, and the Trade and Competitiveness Act of 1988. Under these laws, the US government has engaged in eight rounds of GATT negotiations and participated in creating the World Trade Organization (WTO), a permanent international trade organization responsible for governing the conduct of trade relations among its member countries by administrating the UR Agreement and promoting further liberalization of global trade.

5 In addition to multilateral trade negotiations for free and fair trade practices, the USA has adopted a series of trade remedy laws designed to protect domestic industries from unfair foreign competition. The trade remedy laws that have been commonly used by the US government are those concerning safeguards (the escape clause), countervailing duties, antidumping duties, unfair trading practices under Section 301 of the Trade Act of 1974, and Section 22 of the Agricultural Adjustment Act of 1933.

KEY CONCEPTS

Antidumping duty – A duty imposed to offset unfair trading practices by foreign firms that either sell a product at prices below the average production cost of the product or sell the product at prices below prices the exporters charge for comparable products in their home markets.

Bilateral trade negotiation – Trade negotiation between two trading partners.

Countervailing duty – A duty imposed to offset unfair competitive advantage that foreign manufacturers (or exporters) might enjoy over domestic producers as a result of subsidies.

General Agreement on Tariffs and Trade (GATT) – A treaty that came into effect in 1947, to reduce trade barriers and promote free trade among the member countries.

Material injury – Losses incurred by a domestic industry due to unfair trade practices by foreign firms.

Multilateral trade negotiation – Trade negotiation with more than two countries to promote free trade.

Safeguards (the escape clause) – A provision providing relief to domestic firms and workers desiring protection from increased imports stemming from trade negotiations.

Trade remedy laws – Laws designed to protect domestic industries from unfair foreign competition.

World Trade Organization (WTO) – An international organization created in January 1995 to succeed GATT.

QUESTIONS AND TASKS FOR REVIEW

1 What was the main function of the GATT?
2 How does the GATT differ from the WTO?
3 What are the main items negotiated for agricultural goods in the Uruguay Round of GATT negotiations?
4 What is the purpose of US trade remedy laws and the relation with US commercial policy?
5 Explain the main intention and procedures of the safeguard clause, countervailing duties, antidumping duties, and unfair trading practices under Section 301.

SELECTED BIBLIOGRAPHY

Ingersent, K., Rayner, A. and Hine, R. (eds.) 1994: *Agriculture in the Uruguay Round.* New York: St. Martin's Press.

Josling, T., Tangermann, S. and Warley, T. 1996: *Agriculture in the GATT.* New York: St. Martin's Press.

Koo, W. W. and Uhm, I. H. 2000: U.S.–Canada grain disputes. *Minnesota Journal of Global Trade,* 9 (Winter), 103–19.

Krueger, A. (ed.) 1998: *The WTO as an International Organization.* Chicago: The University of Chicago Press.

Sampson, G. (ed.) 2001: *The Role of the World Trade Organization in Global Governance.* New York: United Nations University Press.

World Trade Organization 1999: *WTO Agreement Series.* Geneva: WTO Publications.

CHAPTER 11   Economic Integration

■ **11.0 INTRODUCTION**

There has been an attempt in recent years to achieve trade liberalization through multilateral agreements, such as the GATT or the WTO. But as the number of countries involved in a negotiation increases, so does the complexity of the negotiations. This decreases the likelihood of reaching an agreement that is acceptable to all parties.

Because of the difficulty in achieving compromise among large groups of countries, there have been a number of bilateral and regional free trade agreements that have been established in recent years. Economic integration occurs when two or more countries join together to form a trading area. One example of economic integration is the creation of a regional free trade area (RFTA). Countries enter RFTA agreements in the expectation of economic gains through trade expansion among the member countries.

There are numerous regional trade agreements throughout the world. Large trade areas, such as the European Union, and bilateral trade agreements, such as the Canada–United States Trade Agreement (CUSTA), create trade opportunities between two or more countries. Yet they often divert existing trade through the barriers to trade with other countries. This is referred to as trade diversion.

This chapter presents different forms of economic integration and discusses its effects on trade flows. Students are introduced to the trade diversion and trade creation effects that take place within free trade areas. The chapter also discusses the impacts of regional free trade agreements on agricultural trade.

■ **11.1 FORMS OF INTERNATIONAL ECONOMIC INTEGRATION**

RFTAs can be classified into several groups in terms of the extent of integration: free trade area, customs union, common market, and economic union. Table 11.1 presents

Table 11.1 Various stages of economic integration

Stage of integration	Industrial free trade area	Full free trade area	Customs union	Common market	Economic union
Abolition of tariffs and quotas among members	yes	yes	yes	yes	yes
Common tariff and quota on nonmember countries	no	no	yes	yes	yes
Abolition of restrictions on factor movement	no	no	no	yes	yes
Harmonization and unification of economic policies and institutions	no	no	no	no	yes

various types of economic integration. A free trade area is established when a group of countries abolishes restrictions on trade between the members, but each member country retains its own tariff and quota system on trade with other countries. An industrial free trade area covers only trade in industrial products, whereas a full free trade area includes all products, including agricultural products and services. The free trade area is one of the most common types of regional integration. The North American Free Trade Agreement (NAFTA) and the Canada–United States Free Trade Agreement (CUSTA) are examples of this type of RFTA.

A *customs union* is created when a group of countries forms a free trade area and also establishes common policies on tariffs and quotas with other countries. MERCOSUR, which includes Brazil, Argentina, Paraguay, and Uruguay, and the Andean Pact, which includes Bolivia, Columbia, Ecuador, Peru, and Venezuela, are examples of customs unions.

A *common market* goes one step further than a customs union by allowing movement of factors of production among members. The European Economic Community (EEC) belonged to this category before it was transformed into the European Union (EU).

An *economic union* is the most comprehensive form of economic integration, in which members have unified fiscal and monetary policies in addition to integration under a common market. Economic union requires a single currency and a central bank, a unified fiscal system, and a common economic policy. Under this system, all member countries give up their economic sovereignty while other economic integration results from the elimination of trade restrictions.

Some examples of RFTAs are CUSTA (1989), the Asian Pacific Economic Cooperation (APEC, 1989), the Southern Cone Common Market Agreement (MERCOSUR, 1991),

BOX 11.1 CUSTA AND NAFTA

The Canada–United States Free Trade Agreement came into effect on January 1, 1989. Under this agreement, tariffs between the two countries were gradually eliminated, and nontariff barriers were gradually reduced. By January 1, 1998, all tariffs were eliminated, with a few exceptions for items covered by tariff-rate quotas (TRQs).

The provisions of CUSTA were incorporated into the North American Free Trade Agreement (NAFTA), which was implemented on January 1, 1994. NAFTA eliminates most barriers to trade and investment between the USA, Canada, and Mexico. In addition to the reduction of trade barriers already provided for under CUSTA, NAFTA calls for the elimination of nontariff barriers and the gradual reduction of tariffs between the USA and Mexico. Many tariffs between the USA and Mexico were eliminated immediately, with others being phased out in periods of 5–15 years. Longer transition periods were given for sectors that are sensitive to imports. However, once the 15-year transition period has passed, all trade between the USA and Mexico will be tariff-free. Products that the USA has included in the 15-year phase-out category include orange juice, sugar, peanuts, certain fresh vegetables, and melons. In addition to the long transition period, NAFTA also allows for special safeguards to protect import-sensitive sectors.

Along with the elimination of tariffs, NAFTA includes provisions on sanitary and phytosanitary measures, export subsidies, internal support, and grade and quality standards. The agreement also provides rules regarding investment, services, intellectual property, competition, and government procurement. It also includes a dispute settlement mechanism.

Implementation of CUSTA and NAFTA has resulted in increased trade between the countries. Since CUSTA took effect, the value of US nonagricultural imports from Canada has more than doubled, increasing from $85 billion in 1989 to $200 billion in 2002. US nonagricultural exports to Canada increased from $73 billion in 1989 to $133 billion in 2002. Agricultural trade has more than tripled during the period of this agreement. US agricultural imports from Canada increased from $2.9 billion to $10.4 billion during the 1989–2002 period, while US agricultural exports to Canada increased from $2.2 billion to $9.9 billion.

From 1993 to 2002, US nonagricultural imports from Mexico increased from $36 billion to $129 billion, while nonagricultural exports to Mexico rose from $37 billion to $79 billion. Agricultural trade with Mexico has doubled since the implementation of NAFTA. US agricultural imports from Mexico rose from $2.7 to $5.3 billion during the 1993–2002 period, while agricultural exports to Mexico increased from $3.5 to $7.1 billion.

In addition to the trade impacts of NAFTA, there are several other areas in which the agreement will impact agriculture and rural communities. NAFTA has created investment opportunities and fostered a more fluid capital market among the countries. In addition, the concern that US jobs would be lost to Mexico did not materialize. While this regional free trade agreement has created incentives for a more efficient allocation of capital, constraints to the movement of labor still remain.

One irony of NAFTA involves the increased amount of trade conflicts since its inception. Allegations of dumping and the use of countervailing duties have become the norm. Disputes involve commodities such as beef carcasses, fresh vegetables, high-fructose corn syrup, and raw and refined sugar. The increased level of trade that has taken place under NAFTA may foster an environment in which disputes are more likely. However, NAFTA has also created mechanisms by which these disputes can be resolved. There are also several other ways in which disputes can be settled prior to their consideration by a dispute settlement panel.

Table 11.2 US trade with Canada and Mexico in agricultural and nonagricultural products (billions of US dollars)

| | US imports | | | | US exports | | | |
| | Canada | | Mexico | | Canada | | Mexico | |
	Ag	Non-Ag	Ag	Non-Ag	Ag	Non-Ag	Ag	Non-Ag
1989	2.9	85.1	2.3	24.3	2.2	72.8	2.7	21.4
1990	3.1	88.1	2.6	26.9	4.1	74.1	2.5	25.0
1991	3.3	87.6	2.5	28.0	4.4	74.3	2.9	29.3
1992	4.0	94.2	2.3	31.6	4.8	78.4	3.7	35.9
1993	4.6	105.9	2.7	36.0	5.2	86.7	3.5	36.8
1994	5.2	123.6	2.8	45.8	5.4	98.2	4.5	44.7
1995	5.5	139.3	3.7	58.0	5.6	107.6	3.4	41.4
1996	6.7	149.6	3.7	70.5	6.0	113.1	5.3	49.3
1997	7.4	160.5	4.0	81.0	6.6	128.2	5.1	63.3
1998	7.7	167.0	4.6	88.5	6.8	131.0	6.0	69.3
1999	7.9	190.4	4.7	104.3	6.8	138.9	5.5	75.9
2000	8.7	220.3	5.0	129.7	7.4	148.2	6.4	94.0
2001	10.0	206.9	5.2	125.3	9.8	134.8	7.3	83.2
2002	10.4	200.1	5.3	128.8	9.9	132.7	7.1	78.9

Source: Interactive Tariff and Trade Dataweb, US International Trade Commission

the Central European Free Trade Agreement (CEFTA, 1992), and NAFTA (1994). APEC is a regional trade initiative, but not a formal RFTA. In the past, RFTAs seldom included agricultural goods in their free trade provisions, or included them only in limited ways. However, the recent RFTAs include agricultural goods.

BOX 11.2 THE FREE TRADE AREA OF THE AMERICAS (FTAA)

Negotiations to create the largest market in the world, the free trade area of the Americas (FTAA), are in progress. The goal of the FTAA is to progressively eliminate trade and investment barriers within the Western Hemisphere. The FTAA will create a market with 800 million consumers and an aggregate GDP of $13 trillion.

The purpose of creating the FTAA is (1) to stimulate economies in the region by increasing trade volume among the member countries, (2) to increase production efficiency through further specialization in production, and (3) to improve social welfare through lowered prices of goods due to enhanced competition.

During the 1994 Summit of the Americas, which was held in Miami in December, the heads of state of the 34 democratic countries in the Western Hemisphere agreed to construct the FTAA and complete negotiations by 2005. Since the initial Summit of the Americas in 1994, negotiations for the FTAA have continued at six trade ministerial meetings, held from June 1995 to April 2001, at the second Summit of the Americas, held at Santiago in April 1998, and at the third Summit of the Americas, held at Quebec City in April 2001. During the 1998 Santiago Summit of the Americas, nine negotiating groups were established: market access (which includes nonagricultural tariffs and nontariff barriers, rules of origin, customs procedures, standards, and safeguards); agriculture (which includes agricultural tariffs and nontariff barriers, agricultural subsidies and other trade-distorting practices, and sanitary and phytosanitary procedures); services; investment; government procurement; intellectual property; subsidies, antidumping, and countervailing duties; competition policy; and dispute settlement. At the sixth ministerial meeting and the third Summit of the Americas, deadlines were fixed for the conclusion and implementation of the agreement. Negotiations are to be concluded no later than January 2005, and the agreement is to be implemented no later than December 2005.

The FTAA will consolidate the numerous free trade agreements currently existing in the Western Hemisphere. There are various different regional trade agreements of different types in the Western Hemisphere. These agreements have put nonparticipating countries at a competitive disadvantage. For example, the MERCOSUR trade agreement includes Argentina, Brazil, Paraguay, and Uruguay. Because of this agreement, US exporters face tariff differentials in the MERCOSUR market that favor member suppliers. The FTAA will help US exporters that are currently outsiders in many of the free trade areas.

Table 11.3 shows economic characteristics of countries and regions in the hemisphere. These countries differ from the USA in terms of the size of economy, per capita income and resource endowments. This suggests that there would be more inter-industry trade through production specialization on the basis of the principle of comparative advantage rather than intra-industry trade. The

Table 11.3 Economic characteristics of the FTAA countries/regions, 2000

Country/ region	GDP[a] (billion $)	Population[b] (millions)	Per capita income[a] ($)	Ag Exp[c] (billion $)	Ag Imp[c] (billion $)
USA	9,825	282	35,019	56	42
Canada	717	31	23,335	16	12
Mexico	581	100	5,754	8	10
C. America	67	37	1,820	5	3
Caribbean	51	20	2,478	1	3
S. America	1,255	350	3,589	33	12
Total	12,494	821		119	81

Sources: [a]International Financial Statistics; [b]US Census Bureau; [c]FTAA Hemisphere Trade and Tariff Database

FTAA could have significant effects on US agricultural trade, since the Western Hemisphere includes key markets for US agricultural products and major suppliers of US agricultural imports. The FTAA may be beneficial for US agriculture, because it will expand market opportunities by progressively eliminating tariffs and nontariff barriers. US agriculture could gain from tariff removal, because agricultural tariffs are higher in other Western Hemisphere countries compared to the US tariffs. The USA imports a large quantity of products, such as coffee and bananas, with no tariffs on these commodities. Once tariffs are eliminated or reduced, it is expected that products that previously faced higher import barriers will experience faster trade growth. This suggests that US agricultural exports will grow faster than imports due to current differences in US and foreign import barriers. Furthermore, tariffs on agricultural products in the Western Hemisphere tend to be higher than tariffs on other products. This suggests that the FTAA may lead to more substantial increases in US trade in agriculture than in other sectors.

BOX 11.3 THE EVOLUTION OF THE EUROPEAN UNION

The European Union as we now know it began after World War II, when European leaders sought to unite their countries economically and politically. This process began in 1951, when six Western European countries agreed to integrate their coal and steel industries. Belgium, West Germany, Luxembourg, France, Italy, and the Netherlands formed the European Coal and Steel Community (ECSC). The ECSC created a body called the High Authority, which had the power to make decisions about the coal and steel industry in these countries.

A few years later, the six members of the ECSC decided to integrate other sectors of their economies. They signed the Treaty of Rome in 1957, which created a customs union, the European Economic Community (EEC). They agreed to remove trade barriers between the six countries and they established a common external tariff. The EEC also established a common market by allowing free movements of labor, capital, and other factors of production. It worked to allow free movement of labor and capital, to abolish trusts and cartels, and to create joint policies on labor, social welfare, agriculture, transportation, and foreign trade.

The number of members in the European Economic Community expanded from six to nine in 1973, when Denmark, Ireland, and the United Kingdom joined. Membership increased to 12 countries after Greece joined in 1981 and Spain and Portugal joined in 1986.

The Maastricht Treaty of 1992, which was entered into force on November 1, 1993, created a new structure known as the European Union (EU) by adding intergovernmental cooperation to the existing "Community" system. This new structure is political as well as economic. Members of the EU make joint decisions on many matters and have developed common policies in a wide range of fields. There are four main branches of the EU's governing body: the European Commission (formerly the Commission of the European Communities), the Council of the European Union (formerly the Council of Ministers of the European Communities), the European Parliament, and the European Court of Justice. The number of countries in the EU increased to 15 in 1995 when Austria, Finland, and Sweden joined.

The Single Market was formally completed at the end of 1992, although there is still some work to be done in some areas. Under a single market, barriers to trade are removed between the member countries, and goods, services, people, and capital can move freely within the Union. Passport and customs checks were abolished at most of the EU's internal borders during the 1990s.

The EU created the European Monetary Union (EMU) in 1992. The EMU created a single currency to be managed by the European Central Bank. The single currency, known as the euro, was introduced on January 1, 2002. Euro notes and coins are used in 12 of the 15 EU countries (Belgium, Germany, Greece, Spain, France, Ireland, Italy, Luxembourg, The Netherlands, Austria, Portugal, and Finland). The UK, Denmark, and Sweden have not adopted the euro and are not members of the EMU.

Currently, the EU is the largest single market in the world after expanding its membership by adding ten more countries in eastern and central Europe. NAFTA is the second largest single market, with an aggregate GDP of $8.9 trillion and 400 million consumers.

■ 11.2 EFFECTS OF ECONOMIC INTEGRATION: TRADE CREATION AND DIVERSION EFFECTS

RFTAs increase trade volume among the member countries through trade creation and trade diversion effects. This section examines these two trade effects in the case of a customs union, the most common form of RFTA.

The trade impacts of an RFTA are divided into trade creation and trade diversion effects. The trade creation effect is defined as an increase in trade volume through the replacement of domestic products with low-priced imports from trading partners. The trade diversion effect is defined as an increase in trade volume through the replacement of imports from the third countries with low-priced imports from trading partners in the customs union. Under NAFTA, the USA increased imports of fruits and vegetables from Mexico, because these products are cheaper to produce in Mexico than in the USA. This increase in imports of fruits and vegetables from Mexico is known as trade creation. On the other hand, the USA increased imports of textile products from Mexico by shifting its import source from China and India to Mexico. Before NAFTA, China and India were lower-priced suppliers of textile products to the USA than Mexico. As a result, the USA imported these products from China and India. Under NAFTA, since the USA eliminated import duties on textile products imported from Mexico, while maintaining import duties on products from China and India, Mexico became the lower-priced supplier of the products to the USA. This increase in imports of textile products is known as trade diversion. Trade diversion is harmful to nonmember exporters.

A simple model will clarify the meaning of trade creation and diversion. Let us assume that there are three countries capable of producing wheat and televisions. Country A is the high-cost producer of both products, country B is the low-cost producer of televisions, and country C is the low-cost producer of wheat, as shown in Tables 11.4 and 11.5. In the absence of restrictions, country A imports televisions

Table 11.4 Trade creation effects of a customs union on television trade (in US dollars)

	Country A	Country B	Country C
Unit price of television	500	430	450
Price plus A's 20% tariff	500	516	540
Price under a customs union	500	430	540

Table 11.5 Trade diversion effects of a customs union on wheat trade (in US dollars)

	Country A	Country B	Country C
Unit price of wheat	5.00	4.30	3.80
Price plus A's 20% tariff	5.00	5.16	4.56
Price under a customs union	5.00	4.30	4.56

from country B and wheat from country C. Suppose that country A imposes a 20% tariff on imports of televisions and wheat. With the tariff in place, country A produces televisions mainly because the prices of televisions from countries B and C ($516 and $540) are higher than country A's price. However, country A still imports wheat from country C because country C's price of wheat ($4.56) is lower than country A's price. Now suppose that countries A and B form a customs union, leaving country C outside the union as a third country. Country A now imports televisions from country B because the price of country B's televisions ($430) is lower than that of televisions produced in country A ($500) under the FTA. These imports from country B represent trade creation.

In the case of wheat, country C is a low-cost producer even with an import tariff of 20%, and country A imports wheat from country C. However, after formation of an FTA between countries A and B, country A shifts its imports of wheat from country C to country B, because country B's price of wheat ($4.30) is lower than country C's price including tariffs ($4.56). These imports of wheat from country B represent trade diversion. Since country A switches its imports of wheat from the most efficient supplier, country C, to a less efficient supplier, country B, under the FTA, the agreement reduces global efficiency in producing wheat and is also harmful to country C.

Trade creation and diversion effects of an FTA are illustrated in Figure 11.1. This figure shows the domestic supply and demand schedules of a commodity in a small importing country. Since this country is assumed to be too small to influence the world market, it faces perfectly elastic export supply schedules from its trading partners, countries B and C. These supply schedules are parallel to the x-axis in Figure 11.1. In Figure 11.1(a), country B is a lower priced supplier (P_1^b) than country C (P_1^c); country B is willing to export the commodity at $3.00 per unit, and country C at $3.50. If country A imposes an import tariff of $1.00 per unit, country B's supply schedule shifts upward from P_1^b to P_2^b and country C's supply schedule shifts from P_1^c to P_2^c. Country A produces 40 units and consumes 55 units, and imports 15 units from country B at the domestic price of $4.00 per unit. Suppose that country A creates an FTA with country B by eliminating the import tariffs on imports from country B, while imposing the tariff on imports from country C. Country B's supply schedule shifts downward to P_1^b. Country A imports 45 units of the commodity from country B at $3.00 per unit. Country A increases its imports by 30 units due to the decreased price from $4.00 to $3.00 under the FTA. The increase in trade volume (30 units) is known as trade creation.

As country A's import price decreases from $4.00 per unit to $3.00 per unit, consumer surpluses in country A increase by area $A + B + C + D$, while producer surplus decreases by area A (Figure 11.1). Area C represents a loss in the tariff revenue collected by country A. Thus, the net gain in social welfare resulting from the FTA is the sum of areas B and D in Figure 11.1(a), indicating that country A is better off as a result of the free trade agreement.

Figure 11.1(b) demonstrates the trade diversion effects of a customs union. It is assumed in this case that country C is a lower-priced supplier than country B. With a tariff of $1.00 per unit imposed by country A, Country A imports 15 units of this

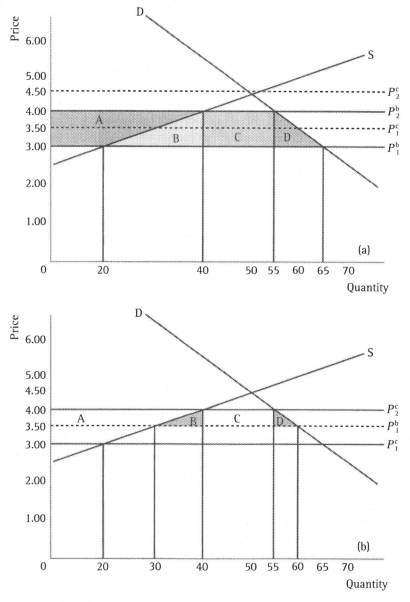

Figure 11.1 The trade creation (a) and trade diversion (b) effects of a customs union.

commodity from country C at the import price of $4.00. At this price, country A produces 40 units and consumes 55 units. If country A forms an FTA with country B and eliminates the tariff of $1.00 on goods imported from country B while maintaining the tariff on goods from country C, country B becomes the lower-priced supplier at $3.50 per unit than country C. As a result, country A switches its import source from country C to country B. Country A produces 30 units and consumes 60 units and imports 30 units at the import price of $3.50 per unit. Country A originally imported 15 units from country C at $4.00 per unit before the FTA, but now imports 30 units from country B after the FTA. The increase in imports (15 units) from country B is known as a trade diversion effect.

As the import price decreases from $4.00 per unit to $3.50 per unit after the FTA, the consumer surplus increases by area $A + B + C + D$, while the producer surplus decreases by area A. Area C represents a loss in the tariff revenue collected by country A. Thus, the net gain in social welfare is the sum of areas B and D.

■ 11.3 IMPLICATIONS OF RFTAS

RFTAs can increase the trade volume among members through trade creation and trade diversion. Trade creation occurs when RFTA members import more from lower-cost RFTA partners. This increases production efficiency among members through production specialization within a region. This also increases consumption efficiency because consumers in an RFTA can buy goods at prices that are lower than those of domestic products. Therefore, the trade creation effects of an RFTA are welfare enhancing.

Trade diversion occurs when members shift imports from more efficient nonmember producers to less efficient member countries. This reduces the global production efficiency. It also distorts optimal trade flows established on the basis of the principle of comparative advantage. Trade diversion, therefore, is harmful to nonmember countries and to the global economy.

Some economists argue that trade diversion is likely to dominate these types of agreements and that the RFTAs will reduce global welfare for nonmember countries. However, others argue that since members already trade with each other and are geographically proximate, trade diversion is minimal, and the RFTA is welfare enhancing. Empirical studies of RFTAs find that aggregate trade creation is much larger than trade diversion and that the RFTAs increase global welfare.

In NAFTA, the trade creation effects are expected to be larger than the trade diversion effects, mainly because Mexico and Canada have been natural trading partners with the USA due to geographical proximity. Prior to the implementation of NAFTA, 75% of Canadian exports and 88% of Mexican exports were destined for the USA. NAFTA tends to enhance production specialization among members and increase efficiency gains in production and consumption.

Another important issue related to the RFTA is its role in global trade liberalization under the WTO. RFTAs may impede the progress of global trade liberalization because of trade diversion effects, suggesting a moratorium on the expansion of RFTAs beyond those already in an advanced stage of negotiation. On the other hand, RFTA expansion may lead to global free trade, since nonmember countries have an incentive to become members as RFTAs expand.

The WTO allows RFTAs with the following restrictions: (1) RFTAs should not be allowed to raise external tariffs against nonmember countries; (2) all barriers to trade between members should be removed; and (3) implementation of the agreement should be completed within a reasonable time frame. The WTO Committee on Regional Trade Agreements (CRTA) was created in 1996 to oversee RFTAs. As long as RFTAs comply with multilateral agreements and reduce external tariffs against nonmember countries, they enhance the process of global trade liberalization through minimizing trade diversion of the RFTAs. Under complete free trade, RFTAs are not needed. On the other hand, countries in an RFTA, especially developing countries, may try to maintain their RFTAs without reducing external barriers. Therefore, they are not likely to participate in the process of multilateral trade liberalization, mainly because they fear competition from the rest of the world. In this case, RFTAs may impede the progress of global trade liberalization.

■ 11.4 CHARACTERISTICS OF TRADE FLOWS UNDER RFTAS

A regional free trade agreement (RFTA) is a free trade agreement among several countries who have similar or dissimilar resource endowments in a region. For example, NAFTA is a free trade agreement among the USA, Canada, and Mexico. Among these countries, the USA and Canada have similar resource endowments, while the resource endowments of Mexico are different from those of the USA and Canada. One question is as follows: What would the effects of the RFTA be on trade flows between the USA and Canada, who have similar resource endowments, and on those between the USA and Mexico, who have different resource endowments?

To answer the question, trade between the USA and Mexico and that between the USA and Canada are analyzed. Since the USA is a more capital- and technology-abundant country than Mexico, while Mexico is a labor-abundant country compared to the USA, an RFTA would increase inter-industry trade between the USA and Mexico. On the basis of the principle of comparative advantage discussed in Chapters 2, 3, and 4, Mexico would have a comparative advantage over the USA in producing labor-intensive commodities, such as leather and textile products, and could increase exports of these products to the USA. The USA has a comparative advantage over Mexico in producing capital-intensive goods, such as computers, automobiles, and aircraft, and could export these goods to Mexico. In addition, on the basis of the differences in climatic conditions between the two countries, Mexico has a comparative advantage over the USA in producing tropical fruits and vegetables. Thus, Mexico exports tropical fruits and vegetables to the USA. On the other hand, again on the basis of climatic conditions and soil types, the USA has a comparative advantage over Mexico in producing grains and oilseeds and can export them to Mexico.

Now look at trade between the USA and Canada under NAFTA. Since both countries have similar resource endowments, trade between the two countries is characterized by more intra-industry trade than inter-industry trade. In spite of similar resource endowments, there has been a significant increase in trade volume between the two countries under NAFTA. The trade between the two countries is not explained by the classical trade theory set out in Chapters 2, 3, and 4, but by the trade theory discussed in Chapter 5.

There are two different models in explaining the causes of increases in trade volume between countries with similar resource endowments when they engage in an RFTA. The first is the monopolistic competition model. Under the assumption that firms produce with increasing returns to scale technology, if countries seek to increase production and marketing efficiency, they do not produce the complete range of products by themselves. The countries may produce a few similar, but differentiated, products to reach external or internal economies of scale and trade them under free trade. Under intra-industry trade, producers benefit because of improved production and marketing efficiency, and consumers are better off because they enjoy an increased number of choices.

The second model is called the national product differentiation model. In this model, products are assumed to be distinguished by country of origin, and the number of varieties supplied by each country is fixed. As a result, goods are different by nationality even though they seem to be similar in nature. Countries increase trade of similar products simply because goods are imperfectly substitutable under free trade.

In the light of the recent trends in trade liberalization, the question of which model can explain the pattern of intra-industry trade in the real world has become important, because these two models predict different effects of trade liberalization. Under the monopolistic competition model, with substantial transportation costs, a higher demand in a larger country attracts foreign firms to locate in the home (large-country) market, so that the larger country serves a smaller market through exports. This is called the "home market" effect, indicating that a larger country has more to gain from the liberalization. By contrast, if goods are distinguished by country of origin, and the number of varieties supplied by each country is fixed, the home market effect is reversed. Under the more liberalized system, the higher level of demand in a larger country encourages its imports of the goods in question. However, there are no new entries to the larger country, so a smaller country gains more from trade liberalization.

SUMMARY

1 Two or more countries form an economic trading area to promote free trade among the member countries by eliminating trade barriers. This is known as a regional free trade area (RFTA).

2 There are several different types of regional free trade areas (RFTAs) in terms of the extent of integration. They are the free trade area, the customs union, the common market, and the economic union.

3 The trade effects of the RFTA are divided into trade creation and trade diversion. The trade creation effects of an RFTA are defined as an increase in trade volume through the replacement of domestic products with low-priced imports from trading partners. The trade diversion effects of an RFTA are defined as an increase in trade volume through the replacement of imports from third-party countries (nonmembers) with low-priced imports from a trading partner with preferential treatment.

4 The trade creation effects of an RFTA increase production efficiency among member countries through production specialization within a region. Consumer efficiency gains occur because consumers in an RFTA can buy goods at prices lower than those of domestic products.

5 The trade diversion effects of an RFTA reduce the global production efficiency, because countries in the RFTA shift their imports from more efficient nonmember countries to less efficient member countries. Thus, an RFTA is harmful for nonmember exporters.

6 Some economists argue that RFTAs impede the progress of global free trade because of their dominating trade diversion effects. However, other economists argue that RFTAs may enhance the progress of global free trade because nonmember countries seek to become members of RFTAs as they expand. The WTO allows RFTAs with certain restrictions, as long as RFTAs do not raise external tariffs against nonmember countries.

7 There would be more inter-industry trade between countries that have dissimilar resource endowments, on the basis of the principle of comparative advantage. On the other hand, intra-industry trade occurs among countries that have similar resource endowments, on the basis of the national product differentiation model and the monopolistic competition model.

KEY CONCEPTS

Canada–United States Free Trade agreement (CUSTA) – A free trade agreement between the USA and Canada.

Common market – The same as customs union, but it allows movement of factors of production among members.

Customs union – A group of countries that abolishes restrictions on trade among the members and also establishes common policies on tariffs and quotas with other countries.

Economic union – The most comprehensive form of economic integration, in which members have unified fiscal and monetary policies in addition to integration under a common market.

European Union (EU) – A regional free trade agreement among several European countries.

Free trade area – A group of countries that abolishes restrictions on trade among the members, who retain their own tariffs and quota systems on trade with other countries.

Inter-industry trade – Exchange of goods between different industries on the basis of the principle of comparative advantage.

Intra-industry trade – Exchange between two countries of similar goods within the same industry.

Monopolistic competition – A type of oligopoly in which firms can differentiate their products from that of their rivals.

National product differentiation – Differentiated products based on the country of origin.

North American Free Trade agreement (NAFTA) – A free trade agreement among the USA, Canada, and Mexico.

Regional free trade area – A treaty with countries in a geographical area to reduce trade barriers and promote free trade.

Trade creation effect – An increase in trade volume through the replacement of domestic products with low-priced imports from trading partners.

Trade diversion effect – A shift in trade volume through the replacement of low-priced imports from third countries with imports from trading partners.

QUESTIONS AND TASKS FOR REVIEW

1   Compare and contrast the different types of regional free trade agreements.
2   What are the economic effects of an RFTA?
3   Is an RFTA good for the world economy? Explain.
4   What are typical characteristics of US trade with Canada and Mexico? If US trade with Canada differs from that with Mexico, why does this occur?
5   Why do some economists argue that RFTAs are harmful to global welfare?
6   Explain intra- and inter-industry trade under RFTAs.
7   How does the EU differ from NAFTA?

SELECTED BIBLIOGRAPHY

Feenstra, R. C., Markusen, J. R. and Rose, A. K. 2001: Using the gravity equation to differentiate among alternative theories of trade. *Canadian Journal of Economics*, 34(2), 430–47.

Head, K. and Ries, J. 2001: Increasing returns versus national product differentiation as an explanation for the pattern of U.S.–Canada trade. *American Economic Review*, 91(4), 858–76.

Helpman, E. and Krugman, P. R. 1985: *Market Structure and Foreign Trade: Increasing Returns, Imperfect Competition, and International Trade*. Cambridge, MA: The MIT Press.

Krugman, P. R. 1980: Scale economies, product differentiation, and the pattern of trade. *American Economic Review*, 70(5), 950–9.

—— 1981: Intraindustry specialization and the gains from trade. *Journal of Political Economy*, 89(5), 959–73.

Linder, S. B. 1961: *An Essay on Trade and Transformation*. Uppsala: Almqvist and Wiksells.

Lipsey, R. G. 1960: The theory of customs union: a general survey. *Economic Journal*, LXX(279), 496–513.

PART
III

Foreign Exchange Markets

CHAPTER 12 | Foreign Exchange Markets and the Exchange Rate

■ 12.0 INTRODUCTION

The quantity of goods and services traded in the world market has increased dramatically in the last 50 years. During the same time period, the amount of international investing has grown, as barriers to financial transactions have decreased. To facilitate international trade and foreign investment, it is important to have an efficient foreign exchange market.

A variety of factors affect commodity markets. Supply and demand work together to determine the equilibrium market price. The foreign exchange market is no different. The willingness of countries, firms, and individuals to buy and sell currency determines the price of currencies on the world market. For example, as the demand for dollars increases it causes the value of the dollar to increase. As the supply of dollars increases, the dollar depreciates. These relationships between supply, demand, and the value of money are critical in understanding the currency exchange market.

A functioning currency exchange market is important for goods and services to be efficiently traded between countries. This chapter examines the functioning of the currency exchange market and its role in determining the exchange rate. The functions of the foreign exchange market are presented and explained, along with an overview of the participants in the market. Important exchange rate systems, specifically the fixed and floating systems, are reviewed. The futures market as a tool for currency trading is also discussed.

This chapter also identifies relationships between the exchange rate and income, prices, and interest rates. On the basis of changes in these factors, the adjustment process and the resulting equilibrium in the foreign exchange market are shown for floating exchange rate systems. Additional specifics regarding the impact of expectations on the exchange rate, the Marshall–Lerner condition, the J-curve effect, and exchange rate overshooting are presented.

■ 12.1 THE CURRENCY EXCHANGE MARKET AND ITS FUNCTIONS

In markets for goods and services, buyers and sellers interact to trade a product. The price of that product is determined by its underlying factors of supply and demand. If consumers desire more of a product, its price will increase; if the supply of that product increases, its price will go down. The interaction of producers and consumers in the market determines the market clearing price for the commodity, the price at which supply equals demand.

THE EFFECTS OF THE EXCHANGE RATE

Consider the example of a commodity such as wheat. Increased production of wheat results in excess supply. This causes the market price to go down, causing consumers to increase their consumption and producers to decrease their production until the market clears. At times when supply does not equal demand, the market price will adjust until the market is once again in equilibrium.

The currency exchange market is no different. It may seem a more difficult concept to grasp, simply because currencies are not products to be consumed. They are a medium of exchange that facilitates the trading of goods and services. Exchange rates are defined as the prices of a currency in terms of other currencies. For example, the exchange rate of US dollars in terms of the Japanese currency might be 120¥ and in terms of the Korean currency might be ₩1,100. Nevertheless, the exchange market and exchange rates impact our economy in a number of ways. For example, the exchange rate affects the ability of companies to export their products and the willingness of consumers to import goods and services. It impacts individuals and governments by influencing the price they must pay to borrow funds and the price they receive on savings deposits. On a more personal level, it directly affects the prices that tourists must pay as they travel to foreign countries. Both producers and consumers are impacted by the exchange rate.

In order to understand how the exchange rate works, suppose the current exchange rate is that 1.50 US dollars equals one British pound ($1.50 = £1.00). Given this exchange rate, an individual holding currency in the amount of £50 could exchange it for an equivalent US dollar amount of $75. Suppose now that the exchange rate changes due to underlying supply and demand conditions in the market. Perhaps demand for dollars increases relative to the pound, resulting in an exchange rate of $1.25 = £1.00. The value of the pound has decreased relative to the dollar. Since $50 \times 1.25 = 62.50$, the £50 note that was initially worth $75.00 is now only worth $62.50. This is referred to as a depreciation of the pound relative to the dollar. From the US perspective the dollar is worth more, since the dollar has appreciated relative to the pound.

Suppose now that the demand for dollars decreases relative to the pound, resulting in an exchange rate of $1.75 = £1.00. The value of the pound has increased relative to the dollar. Since $50 \times 1.75 = 87.50$, the £50 note is now worth $87.50. This is referred to as an appreciation of the pound relative to the dollar, or a depreciation of the dollar relative to the pound.

In either of these situations, the prices of exports and imports are influenced by the exchange rate. As the value of the pound goes up to $1.75 = £1.00$, British consumers can now receive $87.50 worth of goods for £50, as opposed to the initial $75. On the other hand, Americans must now pay $87.50 to purchase £50 worth of British goods, as opposed to the initial $75. Thus, as the value of the British pound increases, US exports to the United Kingdom will increase while UK exports to the United States will decrease. It is clear to see why trade between the two countries is influenced by the exchange rate.

A general rule of thumb is as follows:

1   Given a depreciation of a currency, foreigners find exports cheaper while residents find imports more expensive.
2   Given an appreciation of a currency, foreigners find exports more expensive while residents find imports cheaper.

All else being equal, an appreciation raises the relative price of the country's exports and decreases the relative price of its imports. A depreciation has the opposite effect.

THE FUNCTIONS OF THE CURRENCY EXCHANGE MARKET

Throughout the world, currency is traded by a variety of agents. The *exchange market* is the market in which these international currency trades take place. If the supply of US dollars increases, we could expect the value of the dollar to fall relative to other currencies. Likewise, if the expected return on the US dollar increases, we could expect an increase in the demand for dollars and a corresponding dollar appreciation.

There are a variety of participants in the currency exchange market. The major players are commercial banks, corporations, nonbank financial institutions, and central banks. Individuals can participate in this market, but their role is insignificant relative to that of the major players.

Commercial banks are one of the key players in the currency exchange market, since almost every major transaction involves commercial bank accounts in some capacity. Commercial banks participate in the market in two main capacities. The first involves trades as a service to their customers. Banks exchange currencies in order to facilitate the activities of the businesses and corporations who are their customers. The other capacity involves transactions with other banks. This type of trading, known as inter-bank trading, involves banks buying and selling currencies from each other as they attempt to maximize the return on their money.

Corporations often need to use currency other than that of the country where they are based to pay workers or suppliers. For example, a US firm operating in the UK needs British pounds to pay its workers and local suppliers. This is particularly true in the case of multinational companies with operations in many countries, using a variety of currencies. To accomplish this, corporations utilize the currency exchange market.

Nonbank financial institutions, such as pension funds, mutual funds, and insurance companies become involved in the currency exchange market either as a service to their customers or as a means of investing around the globe.

Central banks often participate in the market as a means of impacting the market, perhaps depreciating the value of its currency to make exports cheaper. While the amount of transactions is typically small, the impact of their participation can be significant given that investors and currency traders typically watch central banks for an indication as to the direction in which the economy and monetary policy are headed. Central banks use intervention to alter the amount of currency in circulation. This affects the balance of payments when countries alter their official reserves. Central bank transaction in the private market for foreign currency assets is known as *official foreign exchange intervention*.

The nature of the currency exchange market has evolved over time. With the advent of electronic trading via computer network systems, trades that previously took days can occur in a matter of seconds. Given this, the integration of financial centers combined with the ability to trade electronically implies no major difference in prices around the world. This phenomenon of buying a currency cheap and selling it for profit is known as *arbitrage*. Discrepancies in prices at any point in the system are eliminated instantaneously as profit-seekers take advantage of market discrepancies. Thus, arbitrage results in an efficient worldwide currency exchange market.

Arbitrage can be seen if there are two separate markets for a currency transactions. Suppose that dollars and pounds can be traded in both the London market and the New York market. If the exchange rate was $1.75 = £1.00 in the London market and $1.50 = £1.00 in the New York market, a trader could buy pounds in the New York market for $1.50 and sell them in the London market for $1.75, making a profit of 25¢ for every pound traded. This opportunity will not last forever. As traders take advantage of this price differential, they increase the demand for pounds in the New York market and the supply of pounds in the London market. This will cause the price of pounds to increase in New York and fall in London. The actions of traders seeking to gain profits through arbitrage will result in an equilibrium dollar–pound exchange rate in both markets.

Another characteristic of the currency exchange market involves the use of *vehicle currencies*. A vehicle currency is used to facilitate currency transactions involving other currencies. For example, an individual wishing to trade Honduran lempiras for Indian rupees may have a difficult time finding someone willing to trade. However, the process may become easier if a vehicle currency, such as the US dollar, is used. Lempiras can be traded for dollars, which are then traded for rupees. Using the dollar as a vehicle currency, the desired outcome is achieved. One qualification for a vehicle currency is that it be widely traded. Examples of common vehicle currencies include the British pound, the euro, the Japanese yen, and the US dollar.

■ 12.2 FORWARD EXCHANGE, CURRENCY EXCHANGE RISK, HEDGING, AND SPECULATION

The currency exchange transactions discussed to this point have taken place on the spot. In other words, an agreement is made between parties for the immediate exchange of currencies for a specified price. The spot market involves the trading of currencies for current or immediate delivery. Given the logistics involved with these types of transactions, immediate delivery may take several days. The price of this transaction is called the *spot exchange rate*.

Currency exchange transactions in the spot market are finalized in a relatively short period of time. However, there is a market that fills a niche for transactions to be executed at some specified time in the future. These forward exchange transactions allow currency traders to specify a currency exchange rate 30 days, 60 days, 90 days, or even a year in the future. These rates are known as *forward exchange rates*. This forward exchange market is particularly beneficial to importers or exporters who have arranged contracts denominated in foreign currency for various dates in the future. Through the use of forward exchange contracts, a forward exchange rate can be locked in, thus guaranteeing a specific price in the trader's home currency.

The futures market allows the risk of currency exchange rate volatility to be minimized. Various futures markets exist that allow currency futures contracts to be traded. One of the largest of these is the International Monetary Market (IMM) of the Chicago Mercantile Exchange.

Futures markets do not allow for trading of all currencies. Specific contracts have been established that allow participants to trade futures contracts in a number of dominant currencies over standard time periods. For example, the IMM allows for the trading of contracts for the exchange of US dollars, Australian dollars, British pounds, Canadian dollars, European Union euros, Japanese yen, Mexican pesos, and Swiss francs.

Contracts are established which call for the delivery of a certain quantity of a currency at some future specified date. For example, contracts may be available for delivery in March, June, September, and December. When traders enter into a futures contract, they are purchasing the right either to receive or deliver a specified quantity of a currency at a specified price at a specified future time.

Two types of traders are involved in the market: speculators and hedgers. Speculators assume risk. Speculators take a position in the market in the hope that the market will move in a direction that allows them to profit from the price change. Hedgers avoid risk. Hedgers use the market to lock in a price, guaranteeing the ability to deliver or obtain currency at some set price in the future. Without speculators, hedgers would not be able to use the futures market to avoid risk.

Suppose that a US exporter will be receiving 125,000 euros next March. Using the futures market, the exporter can lock in a dollar price for those euros in the futures market to minimize the risk associated with exchange rate volatility. Since this individual will actually have €125,000 next March, the hedge will involve selling a futures contract for March delivery at a future price, perhaps 0.95 dollars per EU euro (€1.00 = $0.95). By executing this hedge, the exporter locks in a total dollar value of $118,750 (125,000 × 0.95 = 118,750). Selling a futures contract, as in this example, is taking a short position. Owning the currency is said to be taking an opposite long position. The hedger takes an equal and opposite position in the futures market to the position in the cash market.

The other player in the market is the speculator. While the hedger takes a short or long position in order to guarantee a price, the speculator takes either position in the hope that the market will move in a certain direction. Using the previous example, suppose that a speculator sells a futures contract for March delivery at a specified price. The speculator makes money if the value of euros in terms of dollars declines. If this occurs, he or she can buy back a contract at a lower price, canceling out the previous contract and resulting in a profit. Alternatively, he or she can purchase euros on the cash market in order to deliver them at the higher futures contract price.

The speculator, in seeking profit, absorbs the risk and uncertainty in the market. The hedger reduces exposure to price volatility and locks in a price. The hedger loses any upside potential from profitable swings in the market, but also avoids losses associated with downward movements. The hedger and speculator, although serving different roles, play critical parts in the functioning of the currency futures market.

The options market is an additional market in which future currency assets can be hedged. A currency option is a contract to buy or sell a specific amount of currency at a fixed exchange rate. The right to buy currency is referred to as a call option. The right to sell currency is referred to as a put option. The strike price or exercise price is the price at which currencies can be bought or sold.

Suppose that an American firm is importing products from the EU and has agreed to a contract that specifies the goods will be delivered in three months' time, in exchange for payment in euros. Although the contract specifies the price in euros, exchange rate variability results in uncertainty as to what the dollar price will be after three months. To reduce this uncertainty, the importer can buy a call option to obtain the right to purchase euros at a specified price in three months. Suppose that the strike price were €1.00 = $0.98. If the exchange rate moves to €1.00 = $1.10 in three months, the importer can exercise its option and purchase euros for €1.00 = $0.98. If, on the other hand, the price at the end of three months is €1.00 = $0.90, the importer will let the option expire. In this case, the euros can be purchased on the open market at the lower price of €1.00 = $0.90. As opposed to futures contracts, options allow for the elimination of downside risk without forgoing the upside gains.

■ 12.3 EXCHANGE RATE SYSTEMS AND EXCHANGE RATE DETERMINATION

EXCHANGE RATE SYSTEMS

Countries utilize a floating exchange rate system, a fixed exchange rate system, or a hybrid exchange rate system to determine the value of their currency. Given that most major currencies use floating exchange rate systems, greater emphasis will be placed on the role of floating as opposed to fixed or hybrid exchange rates.

Flexible exchange rate systems

In a floating exchange rate regime, the exchange rate is determined daily by supply and demand conditions. Several major currencies, including US dollars, Japanese yen, British pounds, and Canadian dollars, adopted the floating exchange rate system in 1973. Now most currencies use the floating exchange rate system. Under this system, market forces determine the exchange rate at the level that clears the market.

The underlying supply and demand for currencies in the world market determines the exchange rate. Suppose that the US dollar is initially valued at $1.00 = €1.00. If this is the equilibrium exchange rate, it results in market-clearing conditions, where supply equals demand. Suppose, for whatever reason, that investors experience an increase in their willingness to hold and obtain dollars. If the euro–dollar exchange rate is free to adjust, this shift in the demand for dollars will result in an increase in the value of the dollar, perhaps to $1.00 = £1.10. Regimes in which exchange rates are permitted to adjust to an equilibrium, market-clearing level are known as *floating or flexible exchange rate systems*. In this case, *flexible exchange rates*, in which the exchange rate is determined by the free operation of the market, allow the market to adjust to its equilibrium as derived by the underlying supply and demand.

Fixed exchange rate systems

Exchange rates are not always permitted to adjust on the basis of market conditions. *Fixed exchange rate systems* are characterized by central banks determining the currency price and setting it at that level. Suppose, similar to the earlier example, that the US dollar is fixed by the central bank at $1.00 = €1.00. This could, in fact, be the initial equilibrium exchange rate that results in market-clearing conditions. However, suppose that investors now experience an increase in their willingness to hold and obtain dollars. If the pound–dollar exchange rate of $1.00 = €1.00 is not allowed to adjust, demand for dollars will exceed supply. The dollar will be undervalued and there will be an excess demand for dollars in the currency exchange market. The US government buys euros in the foreign exchange market as much as necessary to maintain the predetermined rate. This will continue until supply equals demand at the target exchange rate.

Governments use fixed exchange rate systems to accomplish various goals. For example, an undervalued exchange rate acts as an import tax and an export subsidy. Fixing the exchange rate at an artificially low level promotes domestic industries by encouraging exports and discouraging imports. It can hurt other industries by increasing the price of imported inputs. An overvalued exchange rate has the opposite effect, acting as an import subsidy and an export tax. Fixing the exchange rate at an artificially high level benefits domestic consumers by encouraging imports and discouraging exports, while decreasing the price of imported inputs.

Hybrid exchange rate systems

Various aspects of the *fixed and floating exchange rate systems* have been combined to form a variety of *hybrid exchange rate systems*. Even fixed exchange rate systems usually allow the exchange rate some freedom to float within a predetermined price band around some fixed target price. There are a variety of widely used hybrid exchange rate systems that can be used by countries to manage their exchange rates. Although the mechanics of these systems may vary, these systems typically involve the exchange rate being fixed over the short term. As chronic balance-of-payments deficits or surpluses develop, monetary authorities are responsible for adjusting the exchange rate to its long-run equilibrium.

EXCHANGE RATE SUPPLY AND DEMAND

People, institutions, and governments hold money for a number of reasons. It is both an investment and a medium of exchange. Given this, the decision to hold money is influenced by factors related to investment and consumption. Three primary factors that affect the decision to hold money are income, prices, and interest rates. As the desire to hold one currency increases relative to others, its value increases. This interaction of the supply and demand for currencies determines the exchange rate. Just as the willingness to hold money is influenced by a variety of factors, those same determinants affect the currency exchange rate. In addition to other factors, the demand for currencies depends on the potential return from investing in that currency and the need to buy or sell currency to facilitate trade in goods and services.

The demand for foreign exchange

The foreign exchange rate at any moment in time will depend on the volume of international transactions that require payments in foreign currency. These transactions may be for purchases of goods and services or purchases of financial assets. The amount of foreign exchange demanded is inversely related to the exchange rate. The amount of currency demanded at a high rate is less than the amount demanded at a low rate, provided that other economic factors (commodity prices, interest rates, and real income) remain the same. For example, an appreciation of the dollar versus the Japanese yen makes US goods and services more expensive in Japan and decreases Japanese imports from the USA. This results in decreased demand for US dollars by the Japanese. Similarly, a depreciation of the US dollar makes US goods and services cheaper in Japan. As a result, Japan increases its imports from the USA. This makes necessary an increase in the quantity demanded of US dollars.

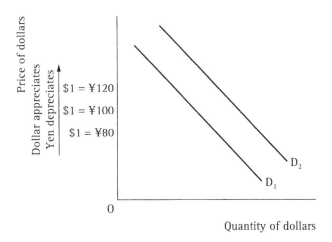

Figure 12.1 Demand for dollars versus yen.

Given this relationship between the exchange rate and foreign exchange demand, the demand for US dollars can be represented by the downward-sloping demand curve shown as curve D_1 in Figure 12.1. As the exchange rate changes, the quantity of dollars demanded will change. This results in a movement along the demand curve D_1. This initial demand for dollars in the foreign exchange market reflects the desire of foreigners to:

- import US goods and services,
- travel to the USA, and
- invest in the USA.

As these factors change, the dollar demand curve can shift. For example, an increase in the desire of foreigners to travel to the USA results in an outward shift in currency demand from D_1 to D_2. This implies that, at any exchange rate, there is an increase in demand for US dollars. Conversely, a decreased desire of foreigners to travel to the USA would result in a decreased demand for US dollars, or a leftward shift in the demand curve.

The supply of foreign exchange

The supply of foreign exchange is based on the level of international transactions that require payment by foreigners. The quantity of foreign exchange supplied to the market varies directly with the rate of exchange. When the value of a foreign currency is high, domestic prices appear low to foreigners. For example, an appreciation of the US dollar makes US goods and services more expensive to the Japanese, while Japanese goods and services are less expensive to US consumers. As a result, US imports of Japanese products increase, resulting in an increase in the quantity

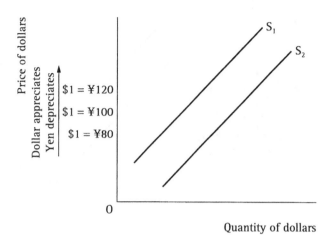

Figure 12.2 Supply of dollars versus yen.

supplied of US dollars. On the other hand, a depreciation of the US dollar makes US goods and services less expensive in the Japanese market. This causes US imports of Japanese products to decrease, resulting in a decrease in the quantity of US dollars supplied.

Similar to the relationship between the exchange rate and foreign exchange demand, the relationship between the exchange rate and foreign exchange supply can be represented by the upward-sloping demand curve shown as curve S_1 in Figure 12.2. As the value of the dollar changes, the quantity of dollars supplied will change. This results in a movement along the supply curve S_1. This initial supply of dollars in the foreign exchange market reflects the desire of Americans to:

- import goods and services,
- travel abroad, and
- invest abroad.

As these factors change the currency supply curve will shift. For example, an increase in the desire of Americans to invest in foreign countries results in an outward shift in currency supply from S_1 to S_2. This implies that, at any exchange rate, there is an increased supply of US dollars. Conversely, a decreased desire of Americans to invest abroad would result in a decreased supply of US dollars, or a leftward shift in the supply curve.

EXCHANGE RATE DETERMINATION IN A FLEXIBLE EXCHANGE RATE SYSTEM

A flexible exchange rate system allows the market to determine the equilibrium exchange rate. Any exchange rate other than the equilibrium rate would result in either excess supply or excess demand. This concept is illustrated in Figure 12.3. The equilibrium

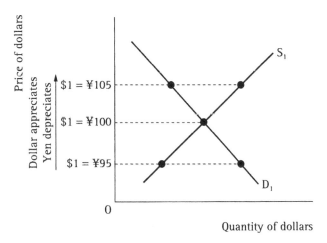

Figure 12.3 Dollar supply and demand.

exchange rate of $1.00 = ¥100 results in demand for dollars being equal to supply of dollars. This situation is in contrast to the exchange rate of $1.00 = ¥95. As the price of dollars goes down, there will be an increase in the demand for dollars and a decrease in the supply of dollars. At the exchange rate of $1.00 = ¥95, there is an excess demand for dollars and the market is in disequilibrium. Similarly, if the exchange rate is $1.00 = ¥105, the supply of dollars will increase while the demand for dollars will decline. At the exchange rate of $1.00 = ¥105, dollars are in excess supply. An equilibrium does not exist in this situation. For supply to equal demand, the exchange rate must adjust to its market-clearing equilibrium. In this case, the market-clearing equilibrium exchange rate is $1.00 = ¥100.

The market-clearing exchange rate is determined by the intersection of the demand and supply schedules. At this exchange rate, the currency market is cleared. That is, the market-clearing exchange rate results in the quantity of currency demanded being equal to the quantity of currency supplied. Once this equilibrium is achieved, the exchange rate will remain stable until a shift occurs in either demand or supply.

As shown in Figure 12.3, the initial equilibrium Japanese yen price of US dollars is $1.00 = ¥100. This market-clearing exchange rate occurs at the point at which the demand for dollars equals the supply of dollars. It will remain at this equilibrium exchange rate until there is a shift either in the dollar supply or the demand curve. The exchange rate continually fluctuates, due to frequent changes in supply of and demand for money.

A currency is strong when the quantity of the currency demanded exceeds the quantity of the currency supplied at a particular exchange rate. This indicates a surplus in the balance of payments. Similarly, a currency is weak when the quantity supplied exceeds the quantity demanded at a particular exchange rate, indicating a balance-of-payments deficit. Government officials in charge of monetary policy are concerned with the size and speed of exchange rate movements. As a result, they

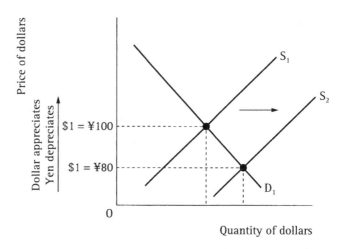

Figure 12.4 The impact of a dollar supply shift.

intervene in the market to stabilize their currency values. This is done through the purchase and sales of foreign currencies in order to stabilize currency values. For example, if the Japanese government desired to stabilize the yen price of US dollars in the foreign exchange market, the Japanese monetary authorities could either purchase or sell yen. Suppose that the Japanese government purchased yen, paying for the yen in dollars on the currency exchange market. As shown in Figure 12.4, this would increase the supply of dollars in the market, causing the supply curve to shift from S_1 to S_2. As a result, the dollar would depreciate while the yen appreciated, from $\$1.00 = ¥100$ to $\$1.00 = ¥80$.

EXCHANGE RATE DETERMINATION IN A FIXED EXCHANGE RATE SYSTEM

Within fixed exchange rate regimes, the exchange rate is set at a predetermined level. Consequently, it cannot always be expected to result in market-clearing conditions. The predetermined exchange rates are maintained through government intervention. Suppose that China fixed the value of Chinese renminbi at $R1.00 = 11.5¢$ and imposed upper and lower limits to the exchange rate, $R1.00 = 12.0¢$ and $R1.00 = 11.0¢$. If the value of the yen with respect to the dollar appreciated to the limit of $R1.00 = 12.0¢$, the Chinese central bank would sell as many renminbi as necessary to decrease the value of the renminbi. Conversely, if the renminbi depreciated to $R1.00 = 11.0¢$, the central bank would buy as many renminbi as necessary to support the value of the renminbi. These transactions would alter the total supply or demand of renminbi relative to dollars, influencing the relative prices of the currencies.

An example of how the government might operate this type of exchange rate regime is shown in Figure 12.5. As the demand for renminbi fluctuates between D_1, D_2, and D_3, its value stays within the predetermined range of $R1.00 = 11.0¢$ and $R1.00 = 12.0¢$. If

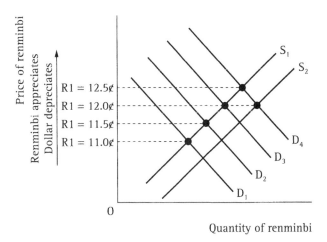

Figure 12.5 Operations of a fixed exchange rate regime.

demand for renminbi were to shift outward to D_4, the value of the renminbi would increase to $R1.00 = 12.5¢$ without government intervention. However, the Chinese government sells renminbi to keep the currency's value within the desired range. This sale of renminbi causes the currency supply to shift outward to S_2. This results in an equilibrium price of $R1.00 = 12.0¢$. The resulting renminbi–dollar exchange rate is now within the bounds set by the government.

■ 12.4 MACROECONOMIC FACTORS AFFECTING THE EXCHANGE RATE

The discussion of variables influencing the supply and demand for money has thus far focused on three general factors. These were the desire to import goods and services, the desire to travel abroad, and the desire to invest abroad. The discussion will now focus on three, more specific, factors that affect the exchange rate. These are real income, interest rates, and prices.

REAL INCOME

Real income is an important factor in the determination of exchange rates as it influences the ability and willingness of countries to purchase goods and services from other countries. If a country's real income (GDP) increases, its residents can afford to spend more. This results in an increase in imports of goods and services, travel, and foreign investment. As this demand changes, so does the demand for foreign currencies. As a result, there is an increase in the supply of the home currency. This causes a depreciation of the home currency.

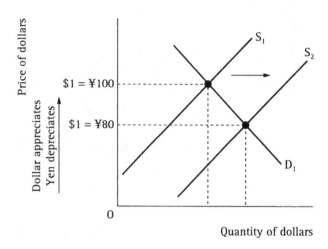

Figure 12.6 Real income and the exchange rate.

Consider the example in Figure 12.6. Given the initial demand and supply, D_1 and S_1, the market-clearing equilibrium occurs at the currency exchange rate of $\$1.00 = ¥100$. Suppose that there is an increase in US real income. This has the effect of increasing US spending on Japanese goods and services which, in turn, leads to an increase in the supply of dollars and a depreciation of the dollar. These effects are represented in Figure 12.6, in which the supply curve shifts to S_2. The resulting market-clearing equilibrium occurs at the currency exchange rate of $\$1.00 = ¥80$.

Similarly, should the Japanese real income decline, Japanese spending on US imports would decrease. As a result, there would be a decrease in the demand for dollars and an appreciation of the yen.

RELATIVE PRICES

As a country experiences inflation, or an increase in its price level relative to other countries, its products in domestic and foreign markets become less competitive. As a result, it imports more foreign goods and services and exports less. An increase in relative prices raises the supply of the home currency, resulting in a depreciation of the home currency and an appreciation of the foreign currency. Conversely, if the home inflation is relatively small compared to other countries, its products are more competitive. This results in a decrease in imports and an increase in exports that, in turn, cause an increase in the value of the home currency.

Consider the interaction between relative prices and the exchange rate as shown in Figure 12.7. As before, the intersection of the initial demand and supply, D_1 and S_1, indicates an initial equilibrium at the currency exchange rate of $\$1.00 = ¥100$. Assume that the US price level rises relative to that of Japan. As a result, US goods and services are less competitive in Japan, while Japanese goods and services become

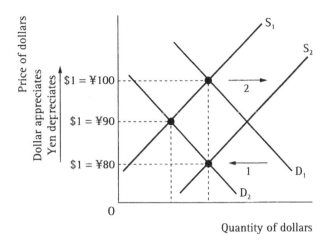

Figure 12.7 Relative prices and the exchange rate.

less expensive in the USA. As Japanese consumers decrease their spending on US goods and services, demand for dollars falls. This decreased demand for dollars is represented by a shift in demand from D_1 to D_2, resulting in a depreciation of the dollar from $\$1.00 = ¥100$ to $\$1.00 = ¥90$. While this has occurred, US consumers now see Japanese products and services as less expensive and increase their demand for Japanese goods and services, resulting in an increased supply of dollars. This increased supply of dollars is represented by the shift from S_1 to S_2, resulting in a further depreciation of the dollar against the yen from $\$1.00 = ¥90$ to $\$1.00 = ¥80$.

RELATIVE INTEREST RATES

If a country's interest rate increases relative to those in other countries, capital will flow into the country with the increased interest rate to take advantage of the higher relative returns. This causes the demand for the country's currency to increase. As a result, this currency appreciates relative to the currencies of other countries. If a country's interest rate decreases relative to other countries, there will be an outflow of capital out of the country and a corresponding decrease in the value of its currency.

An example of the interaction between interest rates and the exchange rate is shown in Figure 12.8. Once again, the intersection of the initial demand and supply, D_1 and S_1, indicates an initial equilibrium at the currency exchange rate of $\$1.00 = ¥100$. Suppose now that the Japanese interest rate falls relative to that of the United States. US citizens will now invest less in Japan, causing the supply of dollars to decrease to S_2. This will cause an appreciation of the dollar to $\$1.00 = ¥110$. However, this is only a portion of the effect. Given the more attractive return on their investment, Japanese citizens will now increase their investments in the USA. This implies an increase in the demand for dollars to D_2. As a result, the impact of this fall in the

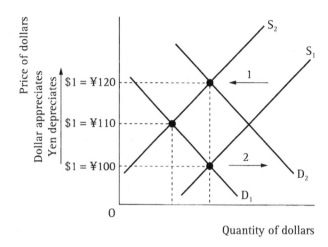

Figure 12.8 Real interest rates and the exchange rate.

Japanese interest rate involves a further appreciation of the US dollar from $1.00 = ¥110 to the new equilibrium of $1.00 = ¥120.

MARKET EXPECTATIONS

It has been shown that exchange rate fluctuations can be caused by movements in several macroeconomic factors. Although real income, relative prices, and interest rates are key determinants of the exchange rate, there are other factors that influence the price of currencies. Exchange rate fluctuations often occur without apparent linkage to market fundamentals. When exchange rates move in a dramatic fashion, they are often reflecting market expectations.

Similar situations occur in other markets. Stock prices often react to news stories and rumor more than price to earnings ratios or other relevant information. The foreign exchange markets work in much the same manner. Expectations and rumors concerning prices, interest rates, imports, exports, productivity, and so on alter willingness to buy or sell currency. In turn, this affects the exchange rate.

Suppose that recent news stories indicate a future lowering of the interest rates in Japan. This affects expectations, causing the yen to depreciate through speculation. An example of the interaction between market expectations and the exchange rate is shown in Figure 12.9. Assume that the intersection of the initial demand and supply, D_1 and S_1, indicates an equilibrium at the currency exchange rate of $1.00 = ¥100. The belief that the yen will depreciate relative to the dollar causes currency traders to purchase dollars for two reasons. Some traders may require a specific quantity of dollars that is sufficient to fill their demand prior to the price increase. Others may simply purchase dollars prior to the yen's depreciation and then sell the dollars later for profit. In either case, this causes the immediate demand for dollars to increase

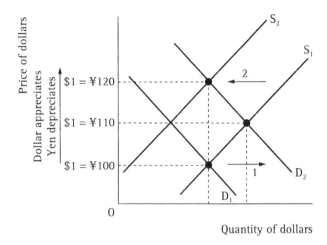

Figure 12.9 Market expectations and the exchange rate.

to D_2. In and of itself, this causes an appreciation of the dollar to $1.00 = ¥110. This increased currency demand increases the price of dollars on its own. However, this is only a portion of the effect.

Those who own dollars and may have been preparing to sell certain quantities may decide to wait until after the expected depreciation of the yen. In other words, they will keep their dollars until the dollar price increases. As a result, dollars are taken off the market and supply shifts to the left. This implies a decrease in the supply of dollars to S_2 and strengthens the depreciation of the yen. As a result, the impact of this expected price increase and interest rate decrease involves a depreciation of the yen from its initial level of $1.00 = ¥100 to the new equilibrium of $1.00 = ¥120.

OTHER FACTORS

To this point, real income, relative prices, interest rates and expectations as factors that influence exchange rates have been discussed. In addition to these determinants, several other factors that influence exchange rate movements will be discussed here.

The *balance of trade* with a particular country will influence the exchange rate. An ongoing US trade deficit with a particular country would create a continual supply of dollars to that country. Over time, investors in that country may become less willing to absorb the inflow of dollars. In this case, the value of the dollar would decline relative to that currency. The opposite would hold true in the case of a trade surplus.

A country's balance of trade is influenced by *consumption patterns*. As consumers' tastes and preferences change, so does their willingness to import. As the desire of consumers to purchase French wine increases, imports of French wine increase, as will the demand for the euro and the supply of the home currency. This results in a depreciation of the home currency relative to the euro. On the other hand, suppose that consumer preferences shift away from French wine and toward domestic varieties. As the desire of consumers to purchase French wine decreases, imports of French wine will decrease, as will the demand for the euro and supply of the home currency. This causes an appreciation of the home currency relative to the euro.

The flow of trade between countries is influenced by both *domestic and trade policy*. Trade restrictions can be used to limit the amount of imports or exports. Export subsidies and domestic support can both encourage exports. Suppose that a country limits imports. This means that less foreign currency is needed to purchase imports and, in turn, less home currency will be supplied to acquire foreign currency. Decreased demand for foreign currency and decreased supply of the home currency causes the home currency to appreciate.

The balance of payments is also influenced by varying levels of *return on investment* between countries. Consider the case in which profitability of investments in Mexico increases relative to the USA. In this case, investors will demand more pesos to invest in the Mexican economy. As investment money moves from the USA to Mexico, this causes the peso to appreciate relative to the dollar.

A factor that affects return on investment is *productivity growth*. Similar to the previous case, suppose that productivity grows in Mexico at a faster rate than in the USA. This is not to say that resources in Mexico are more productive than in the USA, but that productivity is increasing at a faster rate. This results in a decreased relative cost of production, greater profits, and a greater share of the world export market for Mexico. As this occurs, there will be a greater demand for pesos to invest in Mexico and buy its products. This results in an appreciation of the peso.

EXCHANGE RATE OVERSHOOTING

Expectations that affect exchange rates can be influenced through a variety of sources. Because of this, announcements by the government concerning monetary, fiscal, or trade policy are often designed to influence the exchange rate. Information is released to generate excitement and volatility in the foreign exchange market. Individuals modify their expectations and use this information to make decisions that influence the exchange rate. Given that part of the impact comes through expectations, the immediate excitement of the situation may wear off. Changes in market factors then cause a disproportionate short-run impact on exchange rates. In other words, the short-run impact outweighs the long-run effects. This phenomenon is known as *exchange rate overshooting*. An exchange rate is said to overshoot when its impact is greater in the short run than in the long run.

Consider differences in the short-run and long-run supply of dollars, as shown in Figure 12.10. As the demand for dollars increases from D_1 to D_2, the less elastic nature of the short-run supply curve, S_S, results in an initial depreciation of the yen to $1.00 = ¥120. In the long run, the supply of dollars is more elastic. The long-run supply curve, S_L, results in a less severe depreciation of the yen, reaching an equilibrium

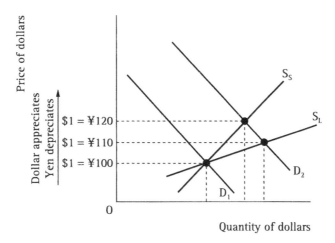

Figure 12.10 Exchange rate overshooting.

exchange rate of $1.00 = ¥110. Given the differences in these elasticities, the short-run exchange rate depreciation overshoots the long-run equilibrium. The initial value of the dollar in terms of yen jumps, but over time it falls back to its long-term equilibrium.

Another explanation for overshooting is that exchange rates tend to be more flexible than other prices. Wages, rental rates, and prices for goods and services tend to be written into contracts, while exchange rates are usually free to fluctuate. As shocks occur in the system, the immediate response occurs in the currency markets. In the short run, the shock to the exchange rate is large, but in the long run other prices can adjust to the shock. This allows the impact of the initial exchange rate shock to diminish. As a result, the short-run impact overshoots the long-run impact that one would expect. But as the effects are able to dissipate, a long-run equilibrium is achieved in the exchange rate.

■ 12.5 THE EXCHANGE RATE AND THE BALANCE OF TRADE

The *balance of trade* is the difference between exports and imports. Under a deficit in the balance of trade, with imports greater than exports, a country must borrow or spend its wealth. On the other hand, when a country has a balance-of-trade surplus, it becomes a net lender or acquires wealth.

Both of these situations show how the balance of trade affects the demand for money. The trade balance of countries ultimately determines relative prices of world currencies. In other words, changes in domestic income and spending influence currency exchange rates. At the same time, changes in the exchange rate influence the balance of trade.

The balance of trade and the exchange rate work in tandem to maintain an equilibrium. Suppose that the USA experiences a trade balance deficit. This implies that the value of imports is greater than the value of exports. Remember that the USA needs foreign currency to import goods from other nations, and other nations need dollars to import US goods. Because of the balance-of-trade deficit, the demand for foreign currency is greater than the demand for dollars. This puts downward pressure on the value of the dollar, resulting in a dollar depreciation. As the dollar depreciates, exports will increase and imports will decrease. This results in a reduction in the current trade account deficit. The opposite holds true in the case of a current account surplus.

THE MARSHALL–LERNER CONDITION

Exchange rate adjustments affect the trade balance through changes in the prices of tradable goods and services. The trade balance is the difference between dollar outpayments and dollar inpayments. Dollar outpayments are equal to imports times the dollar price of foreign goods. Dollar inpayments are equal to exports times the dollar price of US goods. When a country experiences a trade balance deficit, adjustments in the exchange rate and relative prices will occur. A devaluation or depreciation might cause exports to increase and imports to decrease, narrowing or eliminating the trade deficit. The country can accomplish this price adjustment by allowing its exchange rate to depreciate, either through market forces or through a formal currency devaluation. The resulting outcome of the depreciation or devaluation depends on the response of imports and exports to price changes. This responsiveness is measured by the price elasticity of demand for the country's imports (e_m) and the price elasticity of demand for the country's exports (e_x).

If the US dollar is being devalued by 30%, the price of US goods decreases by 30% in terms of the foreign currency and US exports increase. However, the dollar price of US goods remains the same. If US exports increase $(e_x > 0)$, the dollar inpayment will increase. On the other hand, the same dollar depreciation causes the dollar price of foreign goods to increase by 30% and imports to decrease. If imports decrease by more than 30% $(e_m > |1|)$, the dollar outpayments will decrease. As a result, the depreciation of the US dollar increases dollar inpayments and reduces dollar outpayments, resulting in improvements in the trade balance. These are two extreme cases. There are, however, general conditions that allow us to determine whether a country's trade balance will improve, remain the same, or worsen as the result of a depreciation or devaluation. The condition for a real depreciation to improve the current account balance is that

$$|e_x| + |e_m| > 1 \tag{12.1}$$

where e_x is a nation's elasticity of foreign demand for products produced in the country and e_m represents the nation's elasticity of import demand for foreign goods. This is known as the *Marshall–Lerner condition*.

The Marshall–Lerner condition is stated as follows:

| If the sum of the devaluing nation's demand elasticity for imports plus the foreign nations' demand elasticity for exports is greater than (less than) one, then a devaluation will improve (worsen) the devaluing nation's trade balance.

This relationship between exchange rate devaluation or depreciation and the balance of trade is important to remember, because a devaluation or depreciation of a currency will not always improve its balance of trade. As shown by the Marshall–Lerner condition, the validity of this assumption depends upon the response of exports and imports to the exchange rate. In general terms, the Marshall–Lerner condition states that a real devaluation or depreciation improves the balance of trade if exports and imports are sufficiently elastic with respect to the real exchange rate.

Consider the case in which the USA has a balance of trade. Exports are equal to imports. For whatever reason, the US dollar depreciates from an initial price of $1.00 = ¥100$ to $1.00 = ¥80$. The impact of this depreciation on its balance of trade depends on the US elasticity of export demand (e_x) and the elasticity of import demand (e_m). Suppose that e_x is 0.80 and e_m is -0.50. Given this, the 20% depreciation of the dollar will increase exports by 16%, while imports will decrease by 10%. Given that exports increase and imports decrease, it can be seen that the US current account surplus has increased, or its deficit has increased. This is consistent with the Marshall–Lerner condition, which states that if the sum of the devaluing nation's demand elasticity for imports plus the foreign nations' demand elasticity for exports is greater than one, then a depreciation will improve the devaluing nation's trade balance.

THE J-CURVE EFFECT

The Marshall–Lerner condition shows that a devaluation does not necessarily result in an improvement in the balance of trade unless export and import demand are sufficiently elastic. Empirical studies have shown that demand elasticities are usually sufficiently elastic to meet the Marshall–Lerner condition. However, a problem associated with this interaction between exchange rate adjustment and the balance of trade involves a time lag from the actual devaluation and its impact on trade. An examination of the evidence shows that following a devaluation, countries often experience a worsening of their trade balance prior to an improvement. One explanation for this phenomenon is the *J-curve effect*.

A currency devaluation impacts a country's balance of trade. Since a devaluation has the immediate effect of changing relative prices, import expenditures and export receipts will be directly affected. A devaluation causes an increase in exports and import prices in terms of the home currency. This results in the quantity of imports falling, according to the elasticity of import demand, due to higher-priced imports. Exports become more competitive on the world market, resulting in increased exports. The devaluing country should experience a strengthening of its balance of trade. A

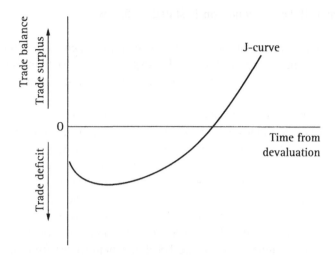

Figure 12.11 The J-curve.

problem associated with this scenario is that it takes time for commodity price adjustments to result in changes in import and export volumes.

The temporal response path of trade flows resulting from a devaluation is illustrated in Figure 12.11. The J-curve is so named because the trade balance worsens for a time until it gradually begins to improve, curving up like a J. The initial response to the devaluation is an increase in import expenditures (dollar outpayments), because contracts made prior to the devaluation must still be honored but their home-currency price is a function of the foreign currency contract price and the new devalued exchange rate. The devaluation, thus, increases the initial price of imports in terms of the home currency without an immediate increase in imports and results in an initial increase in the trade deficit. Over time, quantities adjust to the new price levels. The quantity of imports declines while the quantity of exports increases. The initial devaluation results in an eventual improvement in the balance of trade.

EXCHANGE RATE PASS-THROUGH

Another factor that influences the reaction of the market to an exchange rate movement is *exchange rate pass-through*. As the value of a currency fluctuates, it affects import prices. When a country such as the USA imports a product from the EU, we might assume that a 1% depreciation of the dollar would cause import prices to increase by 1%, resulting in a corresponding decline in imports. This relationship normally holds in the long run. However, over shorter periods of time business may adjust prices to maintain market share or achieve other objectives. The degree of pass-through refers to the import price response to a currency depreciation. If a 1% depreciation results in a 1% increase in import prices, then the degree of pass-through is one, or there is complete pass-through. If a 1% depreciation results in a less than 1% increase in import prices, then there is incomplete pass-through.

Policy-makers often attempt to use a currency devaluation to reduce a trade deficit, but in the presence of incomplete pass-through, the short-term impact is not as great as anticipated. Incomplete pass-through also has important implications for the J-curve phenomenon. Policy-makers should consider these implications as they use devaluation as a tool to influence the trade surplus.

SUMMARY

1 Countries utilize a fixed exchange rate system, a floating exchange rate system, or some combination of the two. Within fixed exchange rate regimes, the exchange rate is set at a predetermined level. Within the floating exchange rate regime, the exchange rate is determined daily by supply and demand conditions.

2 The amount of foreign exchange demanded varies inversely with its price; the amount demanded at a high rate is less than the amount demanded at a low rate, provided that other economic factors (prices, interest rates, and real income) remain the same. Unlike the amount demanded, the quantity of foreign exchange supplied to the market varies directly with the rate of exchange.

3 The market-clearing exchange rate is determined by the intersection of the demand and supply schedules. Once an equilibrium is achieved, the exchange rate will remain stable until a shift occurs in either the demand or supply schedule.

4 The discussion of variables influencing the supply and demand for money focuses on three general factors: (1) the desire to import goods and services; (2) the desire to travel abroad; and (3) the desire to invest abroad. In addition to these three factors, there are other, more specific, factors that affect the exchange rate. These include real income, interest rates, and prices.

5 The *current account balance* is the difference between exports and imports. The Marshall–Lerner condition and the J-curve effect are used to illustrate the interaction between the current account balance and the exchange rate.

6 The *currency exchange market* is the market in which international currency trades take place. It involves interactions between currency buyers and sellers. The price of currencies, or the currency exchange rate, is influenced by the underlying factors of currency supply and demand.

7 When a currency depreciates, it loses value relative to other currencies. Given a depreciation of a currency, foreigners find exports cheaper as residents find imports more expensive. Conversely, when a currency appreciates, it gains value relative to other currencies. Given an appreciation of a currency, foreigners find exports more expensive as residents find imports cheaper.

8 There are a variety of participants in the currency exchange market. The major players are commercial banks, corporations, nonbank financial institutions, and central banks. Individuals can participate in this market, but their role is insignificant relative to that of the major players.

9 Central banks use intervention to alter the amount of currency in circulation. Central bank transactions in the private market for foreign currency assets are known as *official foreign exchange intervention*.

10   With the advent of electronic trading via computer network systems, trades can occur in a matter of seconds. Given this, the integration of financial centers combined with the ability to trade electronically implies no major difference in prices around the world. This phenomenon of buying a currency cheap and selling it for profit is known as *arbitrage.*

11   A *vehicle currency* is a unit of exchange that is used to facilitate currency transactions between other currencies. The dollar, euro, and yen are widely used as vehicle currencies.

12   Regimes in which exchange rates are permitted to adjust to an equilibrium, market-clearing level are known as *floating or flexible exchange rate systems. Fixed exchange rate systems* are characterized by central banks determining the currency price and setting it at that level.

13   The *spot market* involves the trading of currencies for current or immediate delivery. The price of this transaction is called the *spot exchange rate.*

14   The *futures market* fills a niche for transactions to be executed at some specified time in the future. These forward exchange transactions allow currency traders to specify a currency exchange rate 30 days, 60 days, 90 days, or even years in the future. These rates are known as *forward exchange rates.*

15   The *options market* is an additional market that allows future currency assets to be hedged. A currency option is a contract to buy or sell a specific amount of currency at a fixed exchange rate. The right to buy currency is referred to as a call option. The right to sell currency is referred to as a put option.

KEY CONCEPTS

Arbitrage – The phenomenon of buying a currency cheap and selling it for profit.

Currency appreciation – An increase in the value of one currency relative to another.

Currency depreciation – A decrease in the value of one currency relative to another.

Currency exchange market – The forum in which international currency trades take place.

Currency exchange rate – The value of one country's currency in terms of another currency.

Current account balance – The difference between exports and imports.

Exchange rate pass-through – The impact that a depreciation of a currency has on import prices.

Fixed exchange rate systems – The exchange rate regime in which central banks determine the currency price and set it at that level.

Floating or flexible exchange rate systems – The exchange rate regime in which the exchange rate is determined daily by supply and demand conditions.

Forward exchange rates – The price of the transaction involving the trading of currencies for future delivery.

Futures market – The process that allows the risk of currency exchange rate volatility to be transferred from one agent to another.

J-curve effect – The phenomenon in which, following a devaluation, countries often experience a worsening of their trade balance prior to an improvement,

Market-clearing equilibrium – The exchange rate at which currency supply is equal to currency demand.

Marshall–Lerner condition – General conditions that show whether a country's trade balance will improve, remain the same, or worsen as the result of a currency depreciation.

Official foreign exchange intervention – Central bank transactions in the private market for foreign currency assets.

Options – An additional vehicle that allows future currency assets to be hedged.

Spot exchange rate – The price of the transaction involving the trading of currencies for current or immediate delivery.

Spot market – The market in which currency assets are traded for immediate exchange.

Vehicle currency – A unit of exchange that is used to facilitate currency transactions between other currencies.

QUESTIONS AND TASKS FOR REVIEW

1 How is the currency exchange market different than the market for goods and services? How is it the same?

2 Discuss the impact of a currency depreciation on the willingness of consumers to import products. How are exports affected? What happens if the currency appreciates?

3 How can central banks use official foreign exchange intervention to influence the value of their currencies?

4 Discuss how, with the advent of electronic trading through computer network systems, arbitrage results in a more integrated world financial market.

5 Discuss the role of the US dollar as a vehicle currency.

6 How can exporters and importers use the currency futures market to guarantee the forward prices of goods and services to be delivered or received in the future?

7 Define the current account balance. When does a current account surplus exist? When does a current account deficit exist?

8 Show how the current account balance is determined in an open economy.

9 What are the differences between fixed and floating exchange rate systems? How do they differ with respect to the ability of the government to utilize a broad range of monetary and fiscal policies?

10 Show how market factors influence the equilibrium exchange rate in the foreign currency market.

11 What is a hedge? How can a hedge be used to guarantee an exchange rate in the future?

SELECTED BIBLIOGRAPHY

Frankel, J. and Rose, A. 1995: Empirical research on nominal exchange rates. In G. M. Grossman and K. Rogoff (eds.), *Handbook of International Economics*, vol. 3. Amsterdam: North-Holland.

Frenkel, J. and Mussa, M. 1985: Asset markets, exchange rates, and the balance of payments: the reformulation of doctrine. In R. Jones and P. Kenen (eds.), *Handbook of International Economics*, vol. 2. Amsterdam: North-Holland.

Levich, R. 1985: Empirical studies of exchange rates: price behavior, rate determination and market efficiency. In R. Jones and P. Kenen (eds.), *Handbook of International Economics*, vol. 2. Amsterdam: North-Holland.

Obstfeld, M. and Stockman, A. 1985: Exchange rate dynamics. In R. Jones and P. Kenen (eds.), *Handbook of International Economics*, vol. 2. Amsterdam: North-Holland.

CHAPTER 13 | Agricultural Trade and the Exchange Rate

■ 13.0 INTRODUCTION

The current trade balance and the exchange rate are interdependent. Assuming all else equal, an appreciation of the dollar results in decreased US exports and increased US imports. This either lowers the current trade surplus or increases the current account deficit. In either case, the demand for dollars decreases while the supply of dollars increases, resulting in a depreciation of the dollar.

This scenario may hold true for an entire economy. However, a trade surplus or deficit within an individual sector may not be large enough to impact the exchange rate. Agriculture is a relatively small component of the US economy. As a result, changes in agricultural production, consumption, and trade do not have major impacts on the value of the dollar. The exchange rate can be considered an exogenous variable for US agriculture. This may not be the case for developing countries that are more agrarian based. When agriculture accounts for a large share of the economy, it is more likely that an agricultural trade surplus or deficit will affect the exchange rate.

This chapter considers the relationship between the exchange rate and agricultural trade. It presents an analytical framework to assess the impact of currency exchange rate fluctuations. The impact of currency exchange rate fluctuations on agriculture are examined from both a general equilibrium and a partial equilibrium perspective. A general equilibrium perspective is used to analyze the impacts of a deterioration and improvement in the terms of trade. A partial equilibrium framework is then used to provide a more commodity-specific perspective concerning exchange rate impacts on agricultural trade. The partial equilibrium perspective is useful, as it allows the welfare effects of either a depreciation or appreciation to be determined for specific interest groups. Cases of particular commodities and other historical perspectives are used to show the impact of the exchange rate on agricultural trade. Two scenarios are reviewed: (1) exchange rate appreciation for an importing country and exchange rate depreciation for an exporting country; and (2) exchange rate depreciation for an importing country and exchange rate appreciation for an exporting country.

■ 13.1 ANALYZING EXCHANGE RATE IMPACTS: A GENERAL EQUILIBRIUM PERSPECTIVE

Agricultural trade has become increasingly global. As a result, nations have come to the realization that changes in both domestic macroeconomic and trade policies impact the world market. While these policy impacts play a significant role in the trade and welfare of countries, there are other factors at work. The exchange rate is one of these factors.

Changes in the currency exchange rate, or the value of a country's domestic currency in terms of foreign currency, have an immediate impact on the amount of foreign currency that an exporter is willing to receive for its product or that an importer is willing to spend for merchandise. As a result, the impact of changes in domestic and foreign policies can be either compounded or negated by currency exchange rate fluctuations.

BOX 13.1 AGRICULTURAL TRADE AND THE EXCHANGE RATE

US agriculture is highly dependent on the exchange rate. The primary factors determining demand for US agricultural exports are the exchange rate and income in major foreign markets. As agricultural exporters compete to gain access to foreign markets, their competitiveness is determined by the exchange rate, because foreign income affects all exporters. In other words, competition in a foreign market depends on domestic import prices measured in local currencies. Export competitiveness depends on exchange rates between importers and exporters.

Exchange rates affect trade by determining the relationship between international and domestic prices. Changes in the real exchange rate result in the raising or lowering of prices of US goods in local currency around the world. As detailed previously, an appreciation of the dollar raises the price of US goods on the international market, while a depreciation lowers prices. These movements in the exchange rate are particularly important for the agricultural sector in countries such as the USA, where a major portion of agricultural production is exported.

Since it was allowed to float freely in the early 1970s, the dollar has appreciated more with respect to developing country currencies than those of developed countries. This pattern reversed itself in the late 1990s. Beginning in the 1980s, a large number of developing countries have refocused their economies in order to encourage exports and investment. Prior to this, the overvaluation of exchange rates was used by many developing countries to subsidize various industries through their rationing of undervalued foreign exchange.

Fluctuations in currency exchange rates have been shown to account for approximately 25% of the change in US agricultural export value (ERS, 2001). Other variables, such as income growth in developing countries and the growth and productivity of agricultural sectors in competitor countries, accounted for much

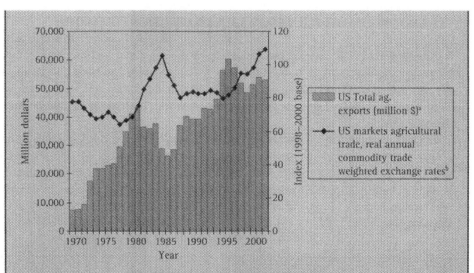

Figure 13.1 Total US agricultural exports versus the commodity-weighted exchange rate. *Source*: ERS/USDA – [a]Annual Commodity Trade Weighted Exchange Rates, www.ers.usda.gov/Data/exchangerates/; [b]Calendar Year US Agricultural Trade, Export Values, www.ers.usda.gov/Data/FATUS

of the rest. However, as indicated in Figure 13.1, the dollar appreciation during the early 1980s and late 1990s corresponded with declines in the value of US agriculture exports. Although other factors contributed to the trade impact, this dollar appreciation allowed competitors to gain market share and expand their production relative to the USA.

Exchange rates are useful in assessing shifts in the competitiveness of products as the value of a currency changes relative to other currencies. Bilateral rates measure the value of a currency versus another particular currency, and are helpful when trying to understand exports to particular markets. The value of a currency becomes even more complicated when considering overall agricultural exports. Complications arise even for a single commodity exported to numerous countries. A measure of value that accounts for a currency's performance against the currencies of competitors and importers is the *trade-weighted exchange rate*. It takes weighted averages of several bilateral exchange rates and combines them into a single index.

Changes in market shares result in variation in trade-weighted exchange rate trends across commodities and commodity groupings. For example, as shown in Figure 13.2, exchange rate patterns for corn, rice, and soybeans have differed due to variations in destination countries. Of the three commodities, rice has shown the greatest appreciation in its trade-weighted exchange rate. This makes US rice more expensive on the world market and has been one factor contributing to the long-term stagnation of US rice exports.

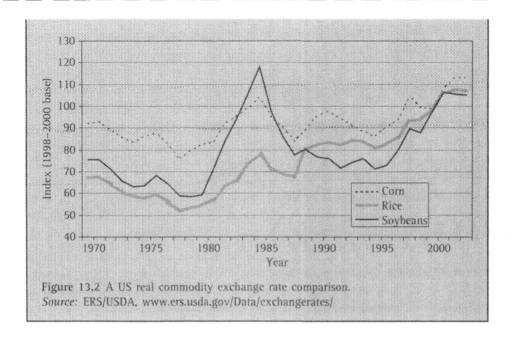

Figure 13.2 A US real commodity exchange rate comparison.
Source: ERS/USDA, www.ers.usda.gov/Data/exchangerates/

For example, consider the impacts of the North American Free Trade Agreement (NAFTA). As Mexico lowered barriers on commodities or products that the United States exports, bilateral trade in those commodities increased between the USA and Mexico since the price that Mexican consumers had to pay for US products decreased. However, shortly after the implementation of the agreement, a major devaluation of the peso occurred. This served to increase the peso price that Mexicans had to pay for US exports. As a result, while trade liberalization served to encourage US exports, changes in the exchange rate negated much of those increases.

Consider the impact of the exchange rate for an economy that produces two goods. This general equilibrium perspective is presented in Figure 13.3. Within this simple framework, the USA can produce and consume some combination of agricultural and manufactured goods. To simplify the analysis, a linear production possibility frontier (PPF) is used. This results in the country specializing in the production of either agricultural or manufactured goods. Given the initial exchange rate, policies, and resulting terms of trade, the USA maximizes its welfare by specializing in the production of agricultural goods. With the initial terms of trade (slope of the income line), TOT_1, the USA produces 18 units of agricultural goods. However, the USA consumes at point C_1 where the income line is tangent to the highest attainable SIC. The USA exports agricultural goods and imports manufactured goods to achieve consumption point C_1. At this point, exports of agricultural goods are eight units $(18 - 10)$ and imports of manufactured goods are 15 units $(15 - 0)$.

Now suppose that the value of the dollar depreciates relative to other currencies. In other words, other countries do not have to use as much of their currency to

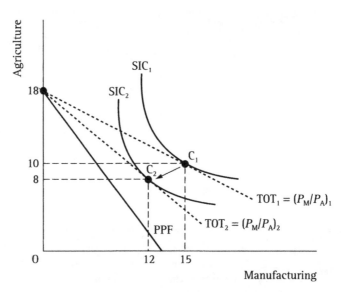

Figure 13.3 Depreciation of the US dollar: a general equilibrium perspective.

purchase one unit of US currency. As a result, other countries find the price of purchasing US agricultural goods cheaper when measured in their own currencies. Those
countries are now willing to increase their imports of agricultural goods from the USA.

On the other hand, as US dollars depreciate relative to other currencies, the USA
will have to use more dollars to purchase the currency of the other countries. As a
result, the USA finds the price of purchasing manufactured imports more expensive
measured in dollars. The USA is now less willing to import manufactured goods from
other countries.

The impacts of this depreciation can be seen in Figure 13.3. As a result of the
depreciation, the terms of trade for the USA rotates from TOT_1 to TOT_2. It is important to note that, in this case, the terms of trade is the ratio of the price of manufactured goods to the price of agricultural goods. In other words, $TOT = P_M/P_A$. In
order to maximize welfare given the new exchange rate, the USA now consumes at
point C_2. At this point, exports of agricultural goods increase to 10 units (18–8) and
imports of manufactured goods decrease to 12 units (12–0). These changes in exports,
imports, and welfare result from changes in price ratios. There is also an income
effect. Although the USA is now exporting more agricultural goods, at the new terms
of trade they are getting less of the manufactured goods. The change in terms of
trade does not cause the country to alter its production from specialization in agricultural goods. By continuing to specialize in agricultural goods as the relative price
of agricultural goods declines, the country is now constrained by a new consumption possibility frontier. Since the agricultural goods are the only source of income
for the USA, the change in the exchange rate and terms of trade ultimately results
in a decrease in both income and welfare for the USA.

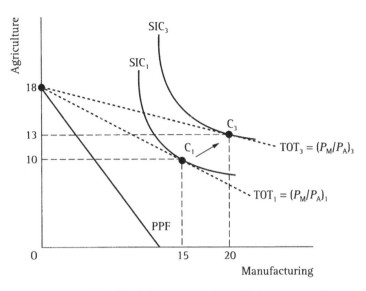

Figure 13.4 Appreciation of the US dollar: a general equilibrium perspective.

An alternative scenario can be examined using Figure 13.4. Suppose that the value of US currency appreciates relative to other countries. In this scenario, other countries must use more of their currency to purchase one unit of US currency. As a result, other countries find the price of purchasing US agricultural goods more expensive when measured in their own currencies. Those countries are less willing to import agricultural goods from the USA.

On the other hand, as US currency appreciates relative to other currencies, the USA will need to use fewer dollars to purchase the currency of other countries. As a result, the USA finds the price of purchasing manufactured goods less expensive as measured in dollars. The USA is now more willing to import manufactured goods from other countries.

The impacts of this currency appreciation can be seen in Figure 13.4. As a result of the currency appreciation, the terms of trade for the USA rotates counter-clockwise from TOT_1 to TOT_3. Maximizing welfare given the new exchange rate, the USA now consumes at point C_3. At this point exports of agricultural goods decrease from eight units (18–10) to five units (18–13) and imports of manufactured goods increase from 15 units (15–0) to 20 units (20–0). As noted with an exchange rate depreciation, there is also an income effect. Although the USA is now exporting less of its agricultural good, at the new terms of trade the country faces a new budget constraint, since it is able to trade one unit of its export good for a larger quantity of imports. As a result, the USA consumes more of the manufactured good. Since the agricultural good is the only source of income for the USA, the change in the exchange rate and terms of trade ultimately results in an increase in both income and welfare for the USA.

BOX 13.2 FLEXIBLE EXCHANGE RATES AND COMPETITIVENESS

Following the collapse of the Bretton Woods system in 1971, the price of the US dollar has floated freely relative to other currencies. As a result, there are as many US dollar exchange rates as there are other currencies. Even countries that fix the nominal price of their currencies in US dollars often have fluctuating inflation-adjusted exchange rates. A currency is a financial asset. Prices of financial assets are usually more volatile than prices of goods. Thus, the incentive to trade varies with the inflation-adjusted exchange rate between countries. However, even countries that do not trade with one another or even compete in common markets can affect each other through their trading partners.

Consider the case of three countries participating in a single commodity market. Suppose that Australia and the USA export beef to Japan. If the bilateral exchange rate between Japan and the USA is considered, an appreciation of the dollar relative to the yen makes US beef more expensive in Japan. Similarly, a depreciation of the dollar reduces the yen price of US beef.

The real world is typically more complicated than this. If the USA competes with Australia for the Japanese market, a depreciation of the dollar may not increase US market share. The important factor is the relative exchange rate. Suppose that the Australian dollar depreciates even more than the US dollar. While US beef has decreased in price, the price of Australian beef has decreased even more. To account for exchange rate variations among customers and competitors, trade-weighted exchange rates are constructed based on the amount that each country trades in a particular market.

A trade-weighted exchange rate is one way to summarize the overall impact of global foreign exchange markets and policies on one country. Since the direct impacts of exchange rates only occur with transactions across international borders, measures of aggregate exchange rates are usually weighted by the value of merchandise trade. Various weighted exchange rate measures, such as the trade weights of the International Monetary Fund (IMF), take into account third-market competition and competition between domestic imports and home production. However, as the IMF points out, no single measure can claim superiority as an indicator of competitiveness.

■ **13.2 ANALYZING EXCHANGE RATE IMPACTS: A PARTIAL EQUILIBRIUM PERSPECTIVE**

While the previous example has been useful to show the broad impact of a change in the value of foreign currency, it fails to show several specific details concerning the workings of the market. This section considers the impact of currency exchange rate fluctuations from a partial equilibrium, commodity perspective. The basic framework is presented in Figure 13.5. Figure 13.5(a) shows domestic demand and supply

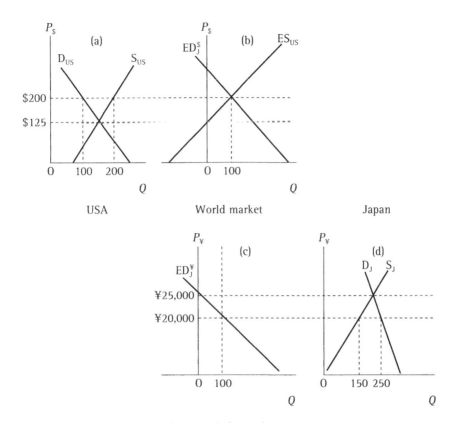

Figure 13.5 A partial equilibrium framework for exchange rate analysis.

schedules in the US market. Assume an exchange rate of ¥100 = $1.00. Within this framework an exporting country, the USA, trades with an importing country, Japan. Given the underlying supply S_{US} and demand D_{US} for the USA, quantity supplied equals quantity demanded at a price of $125. To put it another way, at $125 the export supply of the USA is zero. Given this, the export supply schedule (supply minus demand) for the USA, ES_{US}, is constructed based on the underlying US supply and demand.

As discussed in Chapter 6, Figure 13.5(d) shows the underlying supply S_J and demand D_J for Japan such that supply equals demand at price ¥25,000. At ¥25,000 the export supply of Japan is zero. Given this, the import demand schedule (demand minus supply) for Japan, $ED_J^{¥}$, is constructed based on Japan's underlying supply and demand.

Now consider the currency relationship

$$P_{\$} = P_{¥} \cdot E_{\$/¥}. \tag{13.1}$$

The exchange rate $E_{\$/¥}$ is the value of yen in terms of dollars. Multiplying the yen price by this exchange rate yields the equivalent dollar price. This relationship is used in converting the yen-denominated excess demand curve, $ED_J^{¥}$, in Figure 13.5(c) into the dollar equivalent, $ED_j^\$$, in Figure 13.5(b). As the exchange rate increases in terms of dollars per yen, the excess demand curve, $ED_j^\$$, in Figure 13.5(b) will rotate clockwise due to the increased value of the yen. As the value of the yen declines, the excess demand curve will rotate counter-clockwise given the decreased amount of dollars per yen. This rotation occurs because the change in the exchange rate causes a percentage change in prices at all levels. An example is given in the next section.

Given the export supply and import demand of the two countries, it is now possible to determine the initial level of trade. Suppose that at the initial exchange rate one dollar equals one hundred yen (¥100 = $1.00). Using this exchange rate, the willingness of one country to trade can be shown in terms of the other country's currency. Using Figure 13.5 as an illustration, the yen-based import demand of Japan ($ED_j^\yen$) corresponds with its dollar-equivalent import demand ($ED_j^\$$). Given the export supply and import demand of the two countries shown in a common currency, ES_{US} and $ED_j^\$$, an equilibrium world price is determined at which US exports are equal to Japan's imports. Balanced trade is achieved at a dollar price of $200, where US exports of 100 units equal Japan's imports of 100 units. Converting the dollar price of $200 to its yen equivalent, Japan is willing to import quantity 100 units at a price of ¥20,000.

THE IMPACTS OF A DOLLAR DEPRECIATION

Consider a dollar depreciation. This can be thought of in terms of one yen being worth more dollars, or one dollar buying fewer yen. This example is illustrated in Figure 13.6. Given the initial exchange rate of ¥100 = $1.00, the yen-denominated import demand of Japan, $ED_j^\yen$, corresponds to a dollar-denominated import demand of $ED_j^\$$. In this situation, Japan is willing to import 100 units at a dollar price of $200 or a yen price of ¥20,000.

At the initial exchange rate, machinery worth $200.00 can be purchased for ¥20,000. Consider the situation in which the value of the yen increases. This increase implies that the yen will buy more dollars, or that a dollar will purchase fewer yen. Suppose that this increase in the value of the yen results in an exchange rate of ¥100 = $1.50. As the yen appreciates to ¥100 = $1.50, the ¥20,000 that importers are willing to pay now translates into $300 rather than the previous $200.

It is important to note that the underlying supply and demand of Japan does not change as measured in yen. As a result, the yen-denominated import demand, $ED_j^\yen$, does not change for Japan. Based on the Japanese import demand as shown in Figure 13.6(c), Japan is willing to pay a yen price of ¥15,000 or a dollar price of $150 to import 150 units. However, given the new exchange rate of ¥100 = $1.50, it is now willing to pay a higher price in dollars ($225) to import 150 units while paying the same price in yen (¥15,000).

This willingness to pay a higher price in terms of the foreign currency provides an important illustration regarding changes in the demand for imports as measured in foreign currency. While it has been noted previously that the yen-denominated import demand of Japan does not change, the increased value of the yen results in a

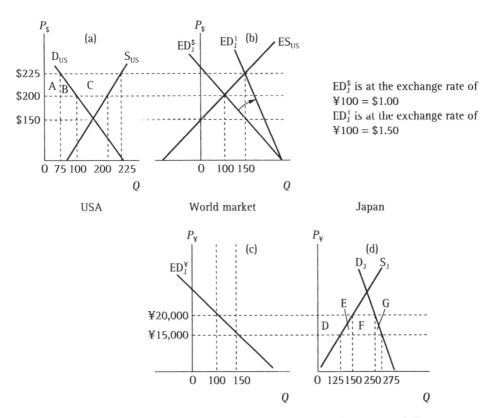

Figure 13.6 Exchange rate appreciation of the Japanese yen against the US dollar.

rotation in Japan's dollar-denominated import demand curve. As shown in Figure 13.6(b), this dollar-denominated import demand curve rotates clockwise from $ED_J^\$$ with the initial exchange rate to ED_J^1 as the value of the yen appreciates.

Note that the nature of this change in the import demand results in a rotation and not a shift in the import demand curve. Given the change in the exchange rate from ¥100 = $1.00 to ¥100 = $1.50, prices represented on the import demand curve ED_J^1 are 1.5 times higher than their corresponding prices on import demand curve $ED_J^\$$. For example, the dollar price ($225) that Japan is willing to pay to import 150 units given the new exchange rate is 1.5 times greater than the dollar price ($150) that Japan was willing to pay to import the same quantity with the previous exchange rate. This process occurs at all points along the import demand curve. As a result, the dollar-denominated import demand curve of Japan pivots on a price of zero (zero times 1.5 equals zero) and rotates clockwise as a result of an appreciation of the yen.

Consider the effects that this depreciation of the dollar has on prices, production, consumption, and trade. Note that no domestic policies have changed; the exchange rate is the only factor that has been modified. On the world market, the rotation of the dollar-denominated import demand curve from $ED_J^\$$ to ED_J^1 results in an increase in the world price from $200 to $225. This price corresponds with an increase in US exports from 100 units to 150 units, while Japan's imports increase from 100 units to 150 units. Since the USA and Japan are the only two countries trading, US exports are equal to Japan's imports at 150 units.

The increase in the world price from $200 to $225 is directly translated into an increase in the US domestic price from $200 to $225. On the basis of this increased price, production in the USA increases to 225 units from its initial level of 200 units. At the same time, consumption decreases to 75 units because of the higher price level. This increase in production and decrease in consumption corresponds with the increase in US exports.

While the increase in the world price from $200 to $225 increases the domestic price in the USA, it has the opposite effect in Japan. Because of the yen appreciation, $225, the new world price in dollars, must be translated into yen to determine Japan's domestic price. This is done simply by finding the price corresponding with 150 units on Japan's import demand curve, $ED_j^\yen$. In Figure 13.6(d), this is shown as a decrease from ¥20,000 to ¥15,000. Although the dollar-denominated price increases, the yen-denominated price decreases. On the basis of this decreased price in Japan, production decreases to 125 units from its initial level of 150 units while consumption increases to 275 units due to the lower price level. This decrease in production and increase in consumption corresponds to the increase in Japan's imports and is identical to the US increase in exports.

The welfare impacts of this exchange rate appreciation for the importing country can be seen in Figure 13.6. Remember that Japanese production decreases from 150 units to 125 units and that consumption increases to 275 units from its initial level of 250 units. These changes are due to a decrease in the domestic price from ¥20,000 to ¥15,000. Given these quantity and price changes, the producer surplus decreases by area D (¥687,500 = ¥5,000 * 137.5). The consumer surplus increases by area D + E + F + G (¥1,312,500 = ¥5,000 * 262.5). This results in a net welfare gain of area E + F + G (¥625,000 = ¥1,312,500 − ¥687,500).

In the case of the USA, the domestic price increases from $200 to $225. This corresponds with a production increase from 200 units to 225 units and a consumption decrease from 100 units to 75 units. The US producer surplus increases by area A + B + C ($5,312.5 = $25 * 212.5). The consumer surplus decreases by area A + B ($2,187.5 = $25 * 87.5). The net welfare effect for the exporting country of a currency depreciation is a gain of area C ($3,125 = $5,312.5 − $2,187.5).

IMPACTS OF A DOLLAR APPRECIATION

Consider a dollar appreciation. Similar to the previous example, this can be thought of in terms of one yen being worth fewer dollars. This example is illustrated in Figure 13.7. Given the initial exchange rate of $1.00 = ¥100, the yen-denominated import demand of Japan, $ED_j^\yen$, corresponds to a dollar-denominated import demand of $ED_j^\$$. In this situation, Japan is willing to import a quantity of 100 units at a price of $200.

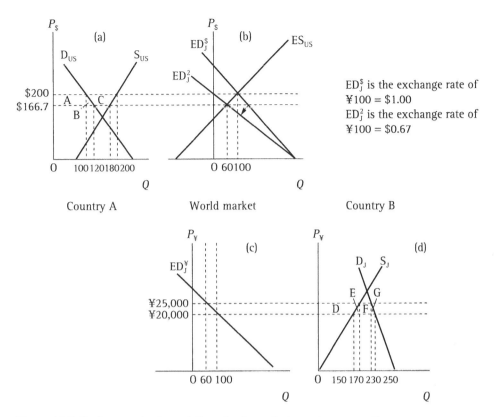

Figure 13.7 Exchange rate depreciation: the importing country perspective.

Suppose now that the yen depreciates. This implies that the yen will buy fewer dollars. Consider the situation in which this increase in the value of the yen results in an exchange rate of ¥100 = $0.67. At the initial exchange rate of ¥100 = $1.00, machinery worth $200 can be purchased for ¥20,000. However, as the dollar appreciates to ¥100 = $0.67, the same machinery worth $200 would cost ¥30,000.

It is important to note once again that the underlying supply and demand of Japan does not change, as measured in yen. As a result, the yen-denominated import demand, $ED_J^¥$, does not change for Japan. Japan was willing to pay a yen price of ¥20,000 or a dollar price of $200 to import 100 units under the initial exchange rate of $1.00 = ¥100. However, given the new exchange rate of ¥100 = $0.67, it is now only willing to pay $133.33, a lower price in dollars, to import 100 units while paying the same price in yen (¥20,000).

This decrease in willingness to pay in terms of the foreign currency provides an important illustration regarding changes in the demand for imports measured in foreign currencies. While it has been noted that the yen-denominated import demand of Japan does not change, the decreased value of the yen results in a rotation in Japan's dollar-denominated import demand curve. As shown in Figure 13.7(b), this dollar-denominated import demand curve rotates counter-clockwise from $ED_J^{\$}$ with the initial exchange rate to ED_J^2 as the value of the yen depreciates.

This change in the import demand results in a rotation and not a shift in the import demand curve. Given the change in the exchange rate from ¥100 = $1.00 to ¥100 = $0.67, prices represented on the import demand curve ED_J^2 are 0.67 times their corresponding prices on import demand curve $ED_J^{\$}$. For example, the dollar price ($133.33) that Japan is willing to pay to import 100 units given the new exchange rate is 0.67 times the dollar price ($200) that Japan was willing to pay to import the same quantity given the previous exchange rate. This process occurs at all points along the demand curve. As a result, the dollar-denominated import demand curve of Japan pivots on a price of zero (zero times 0.67 equals zero) and rotates in a counter-clockwise direction as a result of the depreciation of the yen.

Now consider the effects of this appreciation of the dollar on prices, production, consumption, and trade. Once again, no domestic policies have changed; the exchange rate is the only factor that has been modified. On the world market, the rotation of the dollar-denominated import demand curve from $ED_J^{\$}$ to ED_J^2 results in a decrease in the world price from $200 to $166.7. This price corresponds with a decrease in US exports from 100 units to 60 units, while Japan's exports also decrease from 100 units to 60 units.

The decrease in the world price from $200 to $166.7, since it is denominated in dollars, is directly translated into a decrease in the US domestic price from $200 to $166.7. Based on this decreased price, production in the USA decreases to 180 units from its initial level of 200 units. At the same time, consumption increases to 120 units because of the higher price level. This decrease in production and increase in consumption accounts for the decrease in US exports.

While the decrease in the world price from $200 to $166.7 decreases the domestic price in the USA, it has the opposite effect in Japan. Because of the yen depreciation, the new world price of $166.7 must be translated into yen to determine Japan's domestic price. This is done by finding the price corresponding with the new level of imports on Japan's import demand curve $ED_J^{¥}$. In Figure 13.7 this is shown as a price increase from ¥20,000 to ¥25,000. Although the dollar-denominated price decreases, the yen-denominated price actually increases. Based on this increased domestic price in Japan, production increases to 170 units from its initial level of 150 units, while consumption decreases to 230 units due to the higher price level. This increase in production and decrease in consumption corresponds with the decrease in Japan's imports and is identical to the US decrease in imports.

The welfare impacts of this exchange rate depreciation for the importing country can be seen in Figure 13.7. Remember that Japanese production increases from 150 units to 170 units and that consumption decreases to 230 units from its initial level of 250 units. These changes are due to an increase in the domestic price from ¥20,000 to ¥25,000. Given these quantity and price changes, the producer surplus increases by area D (¥800,000 = ¥5,000 * 160). The consumer surplus decreases by area D + E + F + G (¥1,200,000 = ¥5,000 * 240). This results in a net welfare loss of area E + F + G (¥400,000 = ¥1,200,000 − ¥800,000).

In the case of the USA, the domestic price decreases from $200 to $166.7. This corresponds with a production decrease from 200 units to 180 units and a consumption increase from 100 units to 120 units. The US producer surplus decreases by area A + B + C ($6,327 = $33.3 * 190). The consumer surplus increases by area A + B ($3,663 = $33.3 * 110). The net welfare effect for the exporting country of a currency appreciation is a loss of area C ($2,664 = $6,327 − $3,663).

SUMMARY

1   The exchange rate is one of the factors that can influence the impact of domestic and foreign policy in determining the trade and welfare of an economy. Because changes in the exchange rate have an immediate impact on prices, the impact of changes in domestic and foreign policies can be either compounded or negated by currency exchange rate fluctuations.

2   In the case of an *exchange rate appreciation* for an importing country, the foreign currency price of the product will increase. This results in a lower domestic price for the importer, increased imports, decreased supply, and increased demand. At the same time, the exporter faces a higher domestic price, increased supply, and decreased demand.

3   In the case of an *exchange rate depreciation* for an importing country, the foreign currency price of the product will decrease. This results in a higher domestic price for the importer, decreased imports, increased supply, and decreased demand. At the same time, the exporter faces a lower domestic price, decreased supply, and increased demand.

4   The recent appreciation of the dollar has contributed to decreased US agricultural exports in recent years. From a high of almost $60 billion in 1995, US agricultural exports declined to $49 billion in 1999, a decrease of nearly 20%.

5   *Bilateral exchange rates* measure the value of one currency versus another. However, the value of a currency becomes even more complicated when considering overall agricultural exports. A *trade-weighted exchange rate* accounts for a currency's performance against those of competitors and importers. It takes weighted averages of several bilateral exchange rates and combines them into a single index, based on the importance of a country's various competitors and customers. The trade-weighted exchange rate reflects the exchange rate performance in countries that are important for trade in that commodity, industry, or sector.

KEY CONCEPTS

Bilateral exchange rate – A measure of the value of one currency relative to another currency.

Competitiveness – The ability to deliver a particular quality of goods to consumers at prices equal to or lower than those of competitors.

Currency appreciation – An increase in the value of one currency relative to another.

Currency depreciation – A decrease in the value of one currency relative to another.

Currency devaluation – An action undertaken by the government, designed to decrease the value of its currency relative to other currencies.

Currency fluctuations – Deviations in the value of a currency around its market-clearing equilibrium.

Terms of trade – The price of a country's export commodities relative to the price of its import commodities.

Trade-weighted exchange rate – A measure of value that accounts for a currency's performance against the currencies of competitors and importers.

QUESTIONS AND TASKS FOR REVIEW

1   Explain the impact of an exchange rate depreciation on an economy from a general equilibrium perspective. How is the impact different if the country has a current account surplus versus a current account deficit?

2   Using a partial equilibrium framework, explain the impact of an exchange rate depreciation for an importer. What would the impact be for an exporter?

3   Using a partial equilibrium framework, explain the impact of an exchange rate appreciation for an importer. What would the impact be for an exporter?

4   How would the over-valuation of a country's currency have the same impact as an export tax? How would the same over-valuation have the same impact as an import subsidy?

SELECTED BIBLIOGRAPHY

ERS 2001: Exchange rate indexes & U.S. ag trade. In *Agricultural Outlook*, January–February, AGO-278. USDA, Economic Research Service, Washington, DC.

Orden, D. 2002: Exchange rate effects on agricultural trade. *Journal of Agricultural and Applied Economics*, 34(2), 303–12.

Schuh, E. 1974: The exchange rate and U.S. agriculture. *American Journal of Agricultural Economics*, 56(1), 1–13.

PART IV

Direct Foreign Investment, Economic Growth, and the Environment

PART

IV

Direct Foreign Investment, Economic Growth, and the Environment

International Capital Movements and Multinational Corporations

■ 14.0 INTRODUCTION

International trade involves more than just trade in goods and services. As countries become more integrated, limitations on resource movements between countries are relaxed. The removal of constraints to the flow of capital and labor allows these resources to be better allocated for production. Labor and capital will migrate toward their most profitable use.

So far, commodity trade has been examined assuming that production factors such as capital and labor are immobile. That is, they cannot move between countries. However, factors of production and technology do move across national boundaries. As barriers between countries are relaxed, international commodity trade and movements of productive resources become substitutes. A capital-abundant country, such as the United States, exports capital-intensive commodities or exports capital through foreign investment. It imports labor-intensive goods or allows immigration from labor-abundant countries.

International movements of productive factors have different effects on the nations involved. This chapter will discuss international movements of productive factors, especially the flow of capital through foreign direct investment by multinational corporations (MNCs). The motives of MNCs in deciding whether or not to use foreign direct investment will be considered. In particular, this chapter examines foreign direct investment in the food processing industry.

■ 14.1 MULTINATIONAL CORPORATIONS

Multinational corporations (MNCs) are defined as firms that own, control, or manage production and/or marketing facilities in foreign countries. The number of MNCs has increased since World War II. Today, MNCs produce and handle about 25% of world output. Some examples of MNCs are General Motors, Ford Motors, IBM, Microsoft,

Cargill, and ADM. MNCs expand their operations in foreign countries through foreign direct investment (FDI).

VERTICAL AND HORIZONTAL INTEGRATION

A primary motivation for MNCs is to increase or maintain their cost advantage in producing and distributing commodities in global markets to maximize their profits. MNCs operate through vertical and horizontal integration. Vertical integration occurs when a parent MNC establishes foreign subsidiaries to produce intermediate goods that go into the production of finished products or to produce and market finished goods in host countries. For example, MNCs move their production facilities of labor-intensive goods to labor-abundant countries, such as China and India, to reduce their production costs.

Horizontal integration occurs when a parent company producing a commodity in the home country establishes a subsidiary to produce identical goods produced in host countries. The subsidiaries are established to produce and market the parent company's products in overseas markets. MacDonald's™ and Coca-Cola® are good examples of horizontal integration. They produce homogeneous products in many different countries. Wal-Mart and hotel chains such as Marriott and Sheraton hotels are good examples of horizontal integration in the service industry. They provide similar services in different countries.

MOTIVATIONS OF MULTINATIONAL CORPORATIONS

Most companies establish subsidiaries through vertical and horizontal integration to maintain or enhance their competitive position in producing and distributing their products in global markets. MNCs can ensure the supply of foreign raw materials and intermediate products through vertical integration with companies in foreign countries that produce raw materials or intermediate goods. In addition, MNCs may produce finished goods in foreign countries to make their products more competitive in the markets and circumvent trade restrictions imposed by foreign countries. MNCs can enhance their cost advantages in marketing their products through horizontal integration, allowing them to adapt their products to local conditions.

MNCs may have cost advantage on the basis of economies of scale, development of new technology, and marketing. MNCs may produce in larger volume than independent firms and may be able to increase their competitiveness in global markets through specialization to a greater extent than independent firms. In addition, MNCs devote more resources to research and development of new products. MNCs are in a better position to market their products, to receive information related to a particular product, and to adapt the product quality for regional tastes and preferences.

MNCs establish their subsidiaries to use resources that are much cheaper than in their home countries through vertical integration. For example, MNCs establish their subsidiaries in China to produce labor-intensive goods using Chinese labor, which is much cheaper than labor in their home countries. MNCs are more capable than independent firms in obtaining better trade concessions from foreign governments through negotiations.

■ 14.2 MOVEMENTS OF INTERNATIONAL CAPITAL AND FOREIGN DIRECT INVESTMENT

Foreign direct investment (FDI) refers to investment through ownership of assets in an affiliate by a foreign firm for the purpose of exercising control over the use of those assets. FDI may occur through vertical or horizontal integration with a parent MNC. Most FDI occurs by acquisition, merger, or building new facilities. It is generally known that the country hosting FDI gains from the investing firm's technology, marketing, management, finance, and information services. The gain in employment and economic activity is most obvious from the foreign investment. The parent firm supplying FDI likely improves its acquired firm's production and marketing activities by using the parent firm's technology and information. FDI-supplying firms have some unique assets that give them an advantage over the competing firms in host countries and enable them to generate additional income.

THE INCREASING TREND IN FOREIGN DIRECT INVESTMENT

The USA is one of the leading FDI-supplying countries in the world and one of the leading FDI host countries. As shown in Table 14.1, the annual average for inward and outward flows of FDI into/from the USA amounted to $166 billion and $114 billion, respectively, during the 1995–2000 period. Average FDI outflow outpaces inflow into the USA for the same period. The triad consisting of the USA, Japan, and the European Union dominates FDI, accounting for 60% of FDI inflows and 80% of outflows. FDI outflows outpace inflows in the triad as well as in the world total. Theoretically, the total world inflows of FDI should be equal to the total outflows. However, because of inconsistency in reporting and transaction costs, the total outflows of FDI are shown to be larger than the total inflows of FDI.

EFFECTS OF FOREIGN DIRECT INVESTMENT ON EXPORTS

With respect to the relationship between the FDI and exports by FDI supplying countries, there are two conflicting views on the impact of FDI on the volume of exports to the host countries. They are: (1) that FDI is a substitute for exports originating from the FDI-supplying countries (a substitutive relationship); and (2) that FDI is a complement to exports originating from the FDI suppliers (a complementary relationship).

A substitutive relationship indicates that an increase in FDI will decrease exports to foreign countries and vice versa. In contrast, a complementary relationship indicates

Table 14.1 Foreign direct investment: inward and outward flows ($US billion)

Year	World total		USA		Triad (USA, EU, and Japan)		Triad/world total (%)	
	In[a]	Out[b]	In	Out	In	Out	In	Out
1989–94 (annual average)	200	228	43	49	121	184	61	81
1995	331	355	59	92	172	274	52	77
1996	385	392	84	84	194	290	50	74
1997	478	466	103	96	234	342	49	73
1998	693	712	174	131	438	609	63	86
1999	1,073	1,006	295	143	775	886	72	88
2000	1,271	1,150	281	139	906	943	71	82
1995–2000 (annual average)	706	680	166	114	453	557	60	80

[a] In = inward flows of FDI.
[b] Out = outward flows of FDI.
Source: United Nations Conference on Trade and Development (UNCTAD) 2001: *World Investment Report 2001*. Annex Table B.1: available at http://www.unctade.org/wir/contents/wir01

that FDI and exports move in the same direction. The substitutive relationship originated from the Heckscher–Ohlin theorem, whereby international trade is driven by differences in factor endowments for homogeneous products. These differences in factor endowments become smaller when international factors such as capital become mobile between countries and international trade flows decrease. Thus, capital movements, driven by FDI, are the perfect substitute for exports.

Alternatively, FDI is complementary to exports if FDI outflows create or expand the opportunity to export. The production of one good by foreign affiliates may increase total demand for the entire product line produced by the parent firm supplying FDI, making FDI and exports complementary.

Although this topic has been debated extensively over the last two decades, the issue of the effect of foreign production by a FDI-supplying country's firms on the home country's exports continues to be a puzzle after many years of controversy and a considerable amount of empirical research. In general, empirical studies indicate that the relationship between FDI and exports tends to be substitutive between developed countries and complementary between developed and developing countries.

MOTIVES OF FOREIGN DIRECT INVESTMENT

MNCs carefully consider their decision on FDI on the basis of economic factors, representing profitability in producing a product in either the home country or foreign countries. Other considerations are the host country's rules and regulations related to importing and marketing the products. In general, it is easier to distribute a product produced in the host country than to distribute the same product produced in a foreign country. In addition, MNCs consider social and cultural factors for their decision on FDI.

Economic factors

MNCs seek to maintain or increase profit through reducing production costs by establishing subsidiaries in foreign countries. Such foreign investment may take a number of forms. A firm may establish foreign subsidiaries to ensure a steady supply of essential raw materials necessary to produce a finished good, especially for certain agricultural commodities. If the foreign subsidiary has a natural advantage in producing raw materials, the MNC can reduce its production cost and improve the quality of the finished product.

Another factor that affects decisions on foreign investment is labor costs, which tend to differ between countries. MNCs may be able to reduce their production costs of labor-intensive goods by locating part or all of their productive facilities in foreign countries where wages are lower than those in the home country. Many US firms have subsidiaries in China, mainly because they can reduce their production costs by using cheap Chinese labor.

Transportation costs are another important consideration, especially for an industry in which transportation costs are a high fraction of product value. When transportation costs of raw materials used by MNCs are higher than those of finished goods, MNCs will locate production facilities closer to the foreign country where raw materials are produced than to the market for finished goods. On the other hand, if the transporting costs of raw materials are lower than those of the finished goods, production facilities will be located close to the markets for finished goods.

MNCs may be able to get subsidies from host countries who try to lure foreign manufacturers to invest in their countries. FDI generally stimulates the host country's economy and provides new employment opportunities. In addition, FDI is a way of circumventing import barriers on finished goods. MNCs may be able to avoid high tariffs by producing finished goods in importing countries.

MNCs may make foreign investment in a growing and large market, or in a market that has the potential to be large in the near future. For example, the demand for Western food in China is small, but firms in the food processing industry invest in China on the basis of potential demand for Western foods in the future.

MNCs search for new markets for their finished goods. Some MNCs set up overseas subsidiaries to manufacture and distribute finished goods in overseas markets rather than exporting finished goods directly. This is especially true when familiarity with local conditions in terms of tastes, product design, and packaging is important to successful marketing. MNCs may be able to take advantage by incorporating the local characteristics into their finished goods, to make their products more appealing for consumers in the particular countries.

Another factor considered by MNCs in making FDI decisions is volatility of the value of the host country's currency. If the host country's currency is highly volatile, MNCs are discouraged from investing in the country, mainly because the currency volatility may be a signal of uncertainty of the economy, and MNCs like to avoid uncertainty. On the other hand, depreciation of a currency may attract more FDI if the currency volatility is low.

Social and cultural factors

MNCs may prefer to make production and marketing investment in countries that have a similar culture and use the same language as their home country. In addition, MNCs also prefer to invest in neighboring countries instead of countries far away from the home country because of transportation costs in shipping intermediate or finished goods between home and host countries.

The political environment

Political stability is another factor considered by MNCs when they make decisions on investment in foreign countries. Few MNCs like to invest in politically unstable countries, mainly because of uncertainty associated with their investment stemming from political unrest. Other considerations are the host countries' rules and regulations for exporting and marketing.

■ **14.3 OUTSOURCING AND THE US ECONOMY**

Outsourcing refers to the practice of assigning noncritical parts of a business to another company. When this activity takes place in a foreign country, outsourcing can be viewed as a form of FDI. As discussed in the previous sections, MNCs invest in foreign countries to produce and market goods in those countries. This investment by MNCs is known as FDI. In outsourcing, MNCs own production or service facilities that produce some goods or services at costs lower than in the home country. The product or component is brought back to the home country for consumption or to produce final goods for domestic consumption or exports. Outsourcing is done mainly by MNCs to minimize their production costs in producing a particular line of products. With the advancement of information technology, outsourcing is expanding from assembly lines to the high-technology and service sectors.

Differences in resource endowments between countries are a determinant of outsourcing. For instance, China is a labor-abundant country compared to the USA and wages in China are lower than those in the USA. MNCs in the USA seek to produce components of a labor-intensive product in China to reduce their production costs. The components are brought back to the USA and used to produce a final good to sell in the USA or foreign markets. Given this, outsourcing can be a substitute for imports of labor-intensive goods from China. Without outsourcing, the labor-intensive commodities could simply be imported from China, the labor-abundant country, since it has a comparative advantage over the USA in producing labor-intensive goods. For example, furniture companies in the USA produce over 70% of their wood furniture in China to reduce production costs, since China's wages are much lower than those in the USA. Customer service and computer software engineering jobs are moving to India because Information technology (IT) allows the firms to hire highly educated Indian workers at wages much lower than those in the USA. Workers in the USA have been losing their jobs or suffering wage cuts.

However, outsourcing could enhance global welfare through the efficient utilization of resources in producing goods and providing services. Firms that utilize outsourcing become more competitive in the global market by lowering their production costs. Countries hosting outsourcing increase their national income through economic activities created by outsourcing. Therefore, outsourcing increases global welfare through a more efficient utilization of resources. On the other hand, outsourcing tends to move jobs from the countries that utilize outsourcing to the countries that host outsourcing. For instance, the USA has lost approximately 2.3 million jobs between 2001 and 2003. One reason for this is that MNCs in the USA outsource to India, China, and other countries, including Mexico. In the long run, outsourcing may create new high-technology jobs in the USA, as US firms must become more efficient in order to compete in the global market. Firms can reinvest the additional income for the development of more efficient technologies and additional hiring. The US economy may create enough jobs to offset the losses of jobs to India, China, or other countries.

■ 14.4 FOREIGN DIRECT INVESTMENT IN THE FOOD INDUSTRY

The US food manufacturing industry is not only the largest manufacturing sector in the US economy, accounting for approximately 14% of total US manufacturing output in recent years (i.e., $430 billion), but it also accounts for about one-fourth of the industrialized world's total production of manufactured foods. In addition, trade has been growing faster for manufactured food than for agricultural commodities or intermediate goods. In contrast to global commerce in agricultural commodities, where trade dominates and FDI is insignificant, FDI is the dominant form of international commerce in manufactured foods. Unlike agricultural commodities, manufactured foods are highly differentiated products, with both physical and image differentiation by taste, ingredients, packaging, nutritional value, and advertising.

US food manufacturing firms dominate the list of the world's 50 largest food processing firms. The number of food-manufacturing establishments in the USA reached 20,792 in 1992. The USA accounted for six of the world's 10 largest food-manufacturing firms and 21 of the 50 largest firms. On average, food-manufacturing plants are larger in the USA than in other OECD countries and are more capital intensive. The US share of OECD food processing output in 1992 was 26%, while its share of food processing employment was only about 20%. This indicates that labor productivity in the USA is approximately 30% greater than the OECD average.

Table 14.2 US trade of manufactured food products (SIC-20) ($ billion)

Year	Exports	Imports	Balance
1984	23	17	6
1995	26	18	8
1996	27	21	6
1997	28	23	5
1998	27	24	3
1999	26	26	0
2000	27	27	0
2001	29	28	1

Source: US Department of Agriculture

Table 14.3 The percentage distribution of US exports/imports of manufactured food. Product by product classification, 1999–2001

SIC code	Description	US exports (%)	US imports (%)
201	Meat products	32	15
202	Dairy products	4	5
203	Preserved fruit and vegetables	12	13
204	Grain mill products	14	4
205	Bakery products	2	4
206	Sugar and confections	7	11
207	Fats and oils	13	6
208	Beverages	8	28
209	Fish and miscellaneous	9	13
Total		100.0%	100.0%

Source: US Department of Agriculture

In 1993, international trade in manufactured foods and beverages was valued at $256 billion and accounted for two-thirds of all trade in food and agricultural commodities. US manufactured food exports reached $24 billion in 1993, above the $19 billion in bulk commodity exports. In 1994, sales from the production of foreign affiliates of US food manufacturing firms exceeded $100 billion, more than four times the total value of US exports of manufactured foods. Most of these sales are in foreign markets, with only about 2% being shipped to the USA. Technology, information, communications, and research and development have all played a part in making the USA the leader in the manufactured foods industry. The higher the level of processing, the higher is the level of product differentiation and value added.

The USA has had a trade surplus in manufactured food trade for the last decade. However, the surplus in manufactured food trade has decreased substantially over time due to increased imports, as shown in Table 14.2. The composition of US manufactured food exports is somewhat different from the composition of US imports.

BOX 14.1 US FDI FOR THE PROCESSED FOOD INDUSTRY

Canada and Mexico are the top Western Hemisphere countries for US FDI in
the processed food industry. However, there is also substantial US FDI in South
America. Figure 14.1 shows the US FDI position in Western Hemisphere regions
on a historical cost basis. US FDI in the processed food industry in Canada,
Mexico, and South America has increased substantially since 1989. FDI in South
America and Canada has leveled off since 1996, but FDI in Mexico has continued to grow and has exceeded FDI in Canada. NAFTA may be one contributing
factor for the rapid increases in US FDI in Mexico. FDI in South America is
nearly as large as that in Canada: over half of the FDI in South America is in
Brazil and a large share of the remainder is in Argentina.

Figure 14.1 The US foreign direct investment position abroad on a historical cost
basis in the Western Hemisphere: food and kindred products.

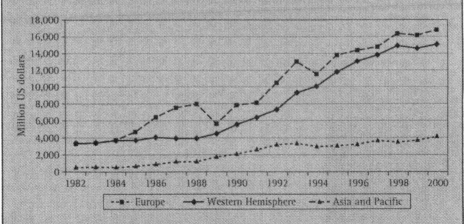

Figure 14.2 The US direct investment position abroad on a historical cost basis: food
and kindred products.

Figure 14.2 compares the US FDI position in the processed foods industry in the Western Hemisphere with that in other parts of the world. US FDI in Europe is slightly higher than that in all of the Western Hemisphere. US FDI in Asia is much lower than in Europe and the Western Hemisphere, due mainly to dissimilarities in culture between the USA and Asia.

These two figures clearly indicate that FDI may be influenced by the host country's economic factors and by cultural linkages between the host and home countries.

Exports of meat products account for the largest share of manufactured food exports (i.e., 32%), followed by grain mill products (14%), as shown in Table 14.3. On the import side, however, the largest share of imports are beverages (28%), followed by meat products (15%). There are some signs of intra-industry trade in manufactured food trade. That is, the USA exports and imports similar classes of goods at the three-digit level of aggregation in Table 14.3. For example, the USA is one of the largest meat-exporting as well as -importing countries.

SUMMARY

1 Multinational corporations (MNCs) are defined as firms that own, control, or manage production and marketing facilities in several countries. The number of MNCs has increased since World War II. The basic reason for their existence is to increase or maintain their competitive position in producing and distributing a commodity in foreign countries.

2 MNCs operate through vertical and horizontal integration. Vertical integration occurs when a parent firm decides to establish foreign subsidiaries to produce intermediate goods that go into the production of finished goods or to market the finished goods of the parent company in the foreign markets. On the other hand, horizontal integration occurs when a parent company producing a commodity sets up a subsidiary to produce identical products in the host country.

3 MNCs establish their subsidiaries for the following reasons: (1) to ensure a steady supply of foreign raw materials or intermediate products; (2) to improve production efficiency through production specialization and the development of new products; and (3) to use production resources that are much cheaper than those in the home country.

4 Foreign direct investment (FDI) refers to investment through ownership of assets in an affiliate by a foreign firm for the purpose of exercising control over the use of those assets. FDI may occur by acquisition or merger or by building new facilities.

5 There are two conflicting views regarding the effects of FDI: (1) that FDI substitutes for exports originating from FDI supplying countries (a substitutive relationship); and (2) that FDI complements exports originating from the FDI suppliers (a complementary relationship).

6 MNCs carefully consider their decision on FDI on the basis of economic factors, social and cultural factors, and the political environment in producing a product in either home or foreign countries. Other considerations are the host countries' rules and regulations for exporting and marketing a product in host countries.

7   Outsourcing refers to the practice of turning over noncritical parts of a business to com-
    panies in foreign countries that specialize in those activities. Outsourcing is done mainly
    by MNCs to minimize their production costs in producing a particular line of products.
    With the advancement of information technology, outsourcing is expanding from assem-
    bly lines to the high-technology and service sectors.

KEY CONCEPTS

Foreign direct investment – Investment through ownership of assets in an affiliate by a foreign
    firm for the purpose of exercising control over the use of those assets.
Foreign subsidiary – A foreign firm which is controlled by a multinational corporation.
Home country – A country providing foreign direct investment.
Horizontal integration – A linkage between a parent company and foreign subsidiaries to pro-
    duce identical goods produced in foreign countries.
Host country – A country receiving foreign direct investment.
Multinational corporation (MNC) – A firm that owns, controls, or manages production **and/or**
    marketing facilities in **foreign** countries.
Outsourcing – The practice of assigning noncritical parts of a business to another company.
Vertical integration – A linkage between a parent firm and foreign subsidiaries to produce
    intermediate goods that go into the production of finished products, or to produce and
    market finished goods in foreign countries.

QUESTIONS AND TASKS FOR REVIEW

1   What is meant by multinational corporations?
2   What are the motives of multinational corporations?
3   What are typical multinational firms operating in the USA and how do they operate?
4   What are the effects of foreign direct investments on the home country's exports?
5   What are motives for foreign direct investment?
6   What is meant by vertical or horizontal integration?
7   Explain economic or cost factors for an MNC's decision on FDI.
8   Explain noneconomic factors for an MNC's decision on FDI.
9   What is outsourcing?
10 How does outsourcing occur and what is the effect of outsourcing on an economy?

SELECTED BIBLIOGRAPHY

Koo, W. W. and Uhm, I. H. 2001: Economic analysis of foreign direct investment and exports:
    a case study of the U.S. food manufacturing industry. *The Journal of World Investment*,
    2(4), 791–804.
Marchant, M. A., Cornell, D. N. and Koo, W. W. 2002: International trade and foreign direct
    investment: substitutes or complements? *Journal of Agricultural and Applied Economics*,
    32(2), 289–302.
Skaksen, M. Y. and Sorensen, J. R. 2001: Should trade unions appreciate foreign direct invest-
    ment? *Journal of International Economics*, 55(2), 379–90.
United Nations Conference on Trade and Development (UNCTAD) 2001: *World Investment Report
    2001*. New York: UN.

CHAPTER 15

Agricultural Trade and Economic Development

■ 15.0 INTRODUCTION

The dynamics of trade negotiations favor rich countries. Countries and trading blocs such as the European Union, Japan, and the United States deal with the trade of large quantities of goods. They make up a large and lucrative share of the world food market. Developed countries are quite interested in increasing their share of these markets.

A logical solution to this issue stems from the stated desire of the developed countries to encourage growth and development throughout the world. The provision of increased market access to less developed country products would serve as a catalyst for growth. It is argued by many that increased market access would do more to stimulate growth than the provision of aid.

Agricultural producers in developed countries make up a relatively small percentage of the population. Despite this, they wield a large amount of political clout and are able to influence both domestic policy and trade policy decisions. Agricultural producers in developing countries often have little political influence and are often dependent on a volatile world market, even though a large percentage of the population belongs to the agricultural sector. These conflicts highlight the dilemma in identifying a politically acceptable level of trade liberalization that will stimulate the desired growth levels in developing countries. World trade in agricultural goods can be liberalized to provide increased market access for less developed countries, thereby stimulating their growth and development.

In considering the potential of achieving this type of environment, this chapter will examine the role of trade in development. Focus will be placed on the increasing role and objectives of developing countries in the World Trade Organization (WTO), the willingness of developed countries to make trade concessions in lieu of financial assistance to spur economic development, and the historic role of food aid programs such as PL480.

■ 15.1 THE ECONOMICS OF AGRICULTURAL TRADE AND DEVELOPMENT

Much of the discussion regarding the ability of a country to produce products thus far has considered endowments, technology, and preferences. A country with a limited amount of labor and capital must decide how to allocate those factors of production to maximize its utility given the available technology. However, as we see in the real world, these underlying factors have a tendency to shift over time. Capital to labor ratios change, new technologies are developed, and the tastes and preferences of individuals change. As a result, comparative advantage also changes over time.

This chapter extends the trade model developed in previous chapters to examine these changes. Shifts in factor endowments and technological advancements are examined to determine how they affect the production possibility frontier (PPF). As shown in Chapter 3, changes in the PPF, combined with potential changes in tastes and preferences, alter the country's offer curve. Such change influences the pattern and volume of trade, and ultimately determines the gains from trade.

ENDOWMENT GROWTH

There are several ways in which an economy can experience growth in its factors of production or endowments. The labor force within a country typically grows in proportion to its population. At the same time, the capital-stock in a country grows as resources are utilized to produce various nonlabor factors of production.

Although there are many different types of capital and labor, to simplify analysis to two factors of production – labor and capital – it is assumed that inputs are homogeneous. We must remember, however, that in the real world there are other factors of production, including natural resources that can be discovered or depleted. This section will assume that the country under analysis is producing two commodities using constant returns to scale technology. The agricultural commodity is capital intensive, while textiles are labor intensive.

An increase in the endowments of labor and capital causes the nation's PPF to shift outward from PPF_1 to PPF_2. The manner in which this shift occurs depends on the rate at which labor (L) and capital (K) grow. If L and K grow at the same rate, the nation's PPF experiences a uniform shift. In this situation, the slopes of both PPFs (p_1 and p_2) are identical at any point along a ray from the origin. This is an example of *balanced growth*. Figure 15.1 presents balanced growth. A proportional increase in labor and capital results in an outward shift in the PPF. Assuming constant prices, the equilibrium production points, which maximize profits under the alternative levels of endowments, are shown as points 1 and 2. Since these production points are on the same ray from the origin (A/T), the proportionate endowment growth has resulted in a proportionate change in production, indicating balanced growth.

If only capital grows, the production of both commodities grows, since capital is used in the production of both commodities. The output of the capital-intensive good,

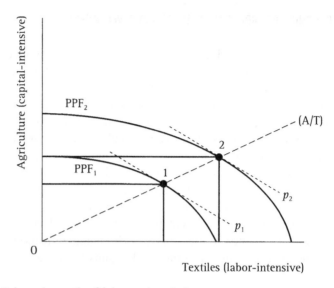

Figure 15.1 Balanced growth of labor and capital.

however, grows more than that of the labor-intensive good. The increase in the endowment of capital causes the production frontier to shift outward from PPF$_1$ to PPF$_2$, as shown in Figure 15.2. But since the growth is due to an increase in capital, the shift in the PPF is skewed toward the production of the capital-intensive good. The equilibrium production point under this new level of endowments with increased capital, assuming constant prices, shifts from point 1 to point 2 in Figure 15.2. Since these production points are not on the same ray from the origin, growth in capital has resulted from a disproportionate change in production. The radius intersecting point 2 is closer to the agriculture axis than the initial radius intersecting point 1. The ratio of agricultural production to textiles production (A/T) has increased from (A/T)$_1$ to (A/T)$_2$. As a result of the increase in capital endowments, this country has shifted its production toward a greater level of specialization in the capital-intensive commodity.

The opposite holds true if the endowment of labor grows. The possible production of both goods increases, but the output of agriculture increases more than that of textiles, as in Figure 15.3. The increase in labor results in the PPF shifting outward, but the shift in the PPF is skewed more toward the production of the labor-intensive good than the capital-intensive good. The equilibrium production point shifts from point 1 to point 3, with a rotation in the agriculture to textiles ratio from (A/T)$_1$ to (A/T)$_3$. Once again, these production points are not on the same ray from the origin, an indication that labor growth has resulted in a disproportionate change in production. The radius intersecting point 3 is closer to the textiles axis than the initial radius intersecting point 1, indicating that the ratio of agricultural production to textiles production (A/T) has decreased. As a result of the increase in labor, this country has shifted its production toward a greater level of specialization in the labor-intensive commodity.

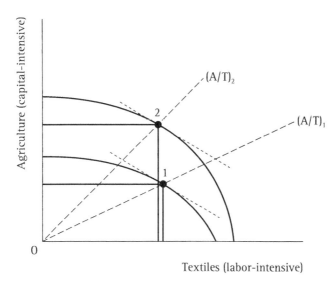

Figure 15.2 Growth in capital greater than growth in labor.

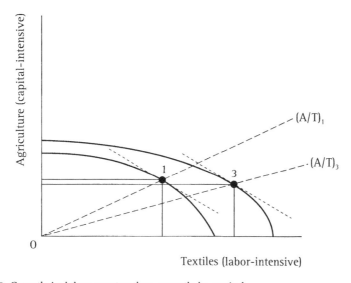

Figure 15.3 Growth in labor greater than growth in capital.

THE RYBCZYNSKI THEOREM

The Rybczynski theorem states that at constant commodity prices, an increase in the endowment of one factor of production increases the output of the good using that factor intensively and reduces the output of the other good. As we have shown in the previous examples, an increase in factor endowments causes the PPF to shift outward. When only one factor is increased, the PPF shift favors production of the good

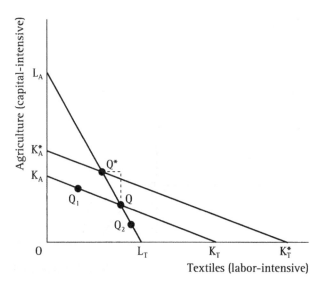

Figure 15.4 An example of the Rybczynski theorem.

that uses that factor most intensively. The Rybczynski theorem goes beyond that. If commodity prices remain constant, the production of the other good will decrease.

Assume that a country is endowed with a given amount of capital ($\bar{K}$) and labor ($\bar{L}$) in producing agricultural and textile products. The equilibrium conditions in the capital and labor markets are

$$\bar{L} = a_x T + a_y A \tag{15.1}$$

$$\bar{K} = b_x T + b_y A \tag{15.2}$$

where a_x and a_y are the units of labor required to produce one unit of agricultural and textile products, respectively, and b_x and b_y are the units of capital required to produce one unit of agricultural and textile products, respectively. The above equations indicate equilibrium conditions in the labor and capital markets.

Labor and capital constraints in the production of agricultural and textile products are shown in Figure 15.4. $L_A L_T$ is a labor constraint when the capital constraint is not limiting. $K_A K_T$ is a capital constraint in an economy when the labor constraint is not limiting. $L_A L_T$ indicates alternative combinations of agricultural goods and textiles that the economy can produce with the given amount of labor (Labor PPF). Similarly, $K_A K_T$ indicates the combination of agricultural goods and textiles that the economy can produce with the given amount of capital (Capital PPF).

When the economy has the given amount of capital and labor, the economy's constraint is $K_A Q L_T$. The economy cannot produce beyond this set of constraints. If the economy produces commodities at point Q_1, capital is fully employed but labor is not. If the economy produces commodities at point Q_2, labor is fully employed but capital is not. When the economy produces at point Q, capital and labor resources are both fully employed.

If the economy initially produces at point Q and moves to a point such as Q_2, the increase in the production of textiles has absorbed all of the labor released by the decreased production of agricultural goods, but has not absorbed all of the capital. This implies that textiles use more labor per unit of capital and are labor intensive. At point Q_1, the economy has absorbed all of the capital released by the decreased production of textiles but not all of the labor. Thus, we can say that agriculture is the capital-intensive commodity.

If the economy increases the supply of capital, the constraint for capital $(K_A K_T)$ shifts outward from $K_A K_T$ to $K_A^* K_T^*$. The full-employment output point is Q^* instead of Q, so agricultural output rises and textiles output falls. This outcome is known as the Rybczynski theorem. The theorem is as follows:

> When factor supplies are fully employed and factor requirements are given, an increase in the supply of one factor of production raises output of the commodity that uses the factor intensively and reduces output of the other commodity.

This theorem is used to show how international differences in factor endowments determine the trade pattern. An algebraic proof of the theorem is presented in the appendix to this chapter.

This is an important point to remember for countries that are trying to increase their welfare through growth. Foreign aid with capital, for example, that may have been intended to increase the overall productivity of an economy, can actually have the effect of drawing labor out of a highly competitive labor-intensive sector of the economy. As a result, the country may produce less of its export goods and more import-competing goods, thus decreasing the level of trade.

■ 15.2 TECHNICAL PROGRESS AND DEVELOPMENT

In addition to the accumulation of capital, the growth that takes place in most industrial countries is due to the development and adoption of more productive technologies. However, analyses of the linkages between growth and technical progress are more complicated than those between growth and endowments, because technical progress can occur at different rates across commodities.

For our purpose, technical progress will be classified as endowment-neutral, labor-biased, or capital-biased. All types of technical progress, regardless of their classification, reduce the inputs required to produce one unit of output. Technical progress includes discoveries that provide improved inputs that allow for increased productivity, as well as the knowledge and ability to better use resources in a manner that improves efficiency. The various types of technical progress are as follows:

- *Endowment-neutral technical progress* increases the productivity of capital and labor in the same proportion. As a result, the capital to labor ratio remains the same as it was prior to the endowment-neutral technical progress, given constant factor prices (wages and rental rates). In other words, there is no substitution between capital and labor, and the only result is that the commodity is now produced with less of both capital and labor.
- *Labor-biased technical progress* increases the productivity of labor to a greater degree than that of capital. As a result, the capital to labor ratio increases, given constant factor prices, because less labor is used relative to capital. In other words, capital is substituted for labor in the production process. Since the per unit share of capital used in the production process increases relative to labor, this type of technical progress is called labor-biased. The commodity can now be produced using relatively less labor, resulting in an increased capital to labor ratio.
- *Capital-biased technical progress* increases the productivity of capital to a greater degree than that of labor. As a result, the capital to labor ratio decreases given constant factor prices, since less capital is used relative to labor. Labor is substituted for capital in the production process. Since the per unit share of labor is used in the production process increases relative to capital, this type of technical progress is called capital-biased. The commodity is now produced using relatively less capital, resulting in a decreased capital to labor ratio.

As with endowment growth, all types of technical progress result in an outward shift in the PPF. This shift depends on the type and degree of technical progress that is occurring in either or both commodities. Since both capital-biased and labor-biased technical progress are difficult to deal with in a nonmathematical manner, only endowment-neutral technical progress is considered in this chapter.

Consider the case of two commodities experiencing endowment-neutral technical progress. If both commodities experience the same level of endowment-neutral technical progress, the PPF shifts outward, similar to the balanced endowment growth in Figure 15.1. Given this uniform shift in the nation's PPF, the slopes of both PPFs are identical at points intersected by a radius from the origin. Assuming constant prices as with balanced endowment growth, equilibrium production points under these alternative levels of technical progress are shown as points 1 and 2 in Figure 15.1. Since these production points are on the same radius from the origin, this identical rate of endowment-neutral technical progress has resulted in a proportionate change in production.

Now suppose that only one of the commodities experiences endowment-neutral technical progress. Assume that endowment-neutral technical progress occurs in the textiles sector and that no technical progress occurs in the agricultural sector. Since no technical progress has occurred in agriculture, each level of input will result in the same level of output as before. In particular, the PPF intersects the vertical axis at the same point in both cases. But since textiles have experienced endowment-neutral technical change, the resulting PPF will shift to the right by some scalar. For

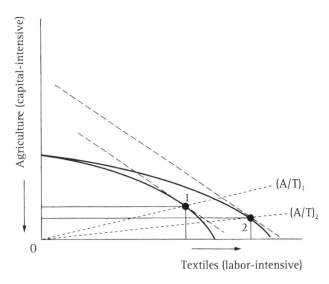

Figure 15.5 The impact of labor-biased technical progress on the PPF.

example, suppose that the productivity of labor and capital increases for textiles, while that for agriculture remains constant. Since factor productivity has increased for textiles, each potential combination of capital and labor results in a rightward shift in the PPF, as shown in Figure 15.5.

The corresponding scenario can occur, in which agriculture is the only sector that experiences endowment-neutral technical progress. Consider Figure 15.6, where endowment-neutral technical progress occurs only in the agricultural sector. Since no technical progress has occurred in textiles, each level of input will result in the same level of output as before. In particular, the PPF intersects the horizontal or textiles axis at the same point. But since the agricultural sector experiences endowment-neutral technical progress, the resulting PPF will shift upward by some scalar.

It is important to note that although both cases involve endowment-neutral technical progress, neither is commodity-neutral. In the case of endowment-neutral technical progress in the textiles sector, the equilibrium production points under these alternative levels of technical progress are shown as points 1 and 2 in Figure 15.5. Since point 2 is on a radius closer to the textiles axis, this endowment-neutral textiles-biased technical progress has resulted in a disproportionate change in production. Since the textiles sector uses labor more intensively than agriculture, this commodity-biased technical progress will cause the demand for labor to increase relative to capital.

The corresponding situation occurs in the agricultural sector. In this case, the equilibrium production points under the initial level of technology and when technical progress occurs in the agricultural sector are shown as points 1 and 3 in Figure 15.6. Since point 3 is on a radius closer to the agricultural axis, this endowment-neutral but agricultural-biased technical progress has resulted in a disproportionate change

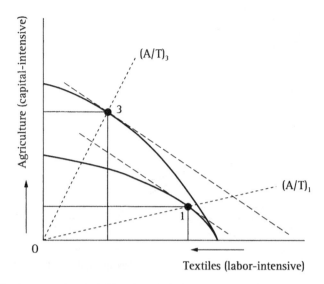

Figure 15.6 The impact of capital-biased technical progress on the PPF.

in production. Since the agricultural sector uses capital more intensively than does the textiles sector, this commodity-biased technical progress will cause the demand for capital to increase relative to labor. This occurs despite the fact that the new technology is endowment-neutral.

■ 15.3 ECONOMIC GROWTH AND TRADE

Consider now the impact of growth on trade. Up to this point, we have examined the effect of growth in factor endowments or technical progress on the PPF. We now extend the analysis to show how shifts in the PPF impact an offer curve and ultimately trade and welfare.

   1 *The small-country scenario.* The trade effects on growth in the case of the small country can be seen in Figure 15.7. The left panel is similar to the example of growth in the endowment of labor as shown in Figure 15.3. In this case, as a small country, prices do not change and the social indifference curves are included in order to show the change in consumption. Given the constant world price of 1 $(P_T/P_A = P = 1)$, this country selects its initial level of production at P_1, which will allow it to consume at point C_1 in the left panel and trade to the corresponding point 1′ in the right panel. As the PPF shifts outward, the country selects the production point P_2 that allows it to consume at point C_2 and trade to the corresponding point 2′.

   Production of the export good increases while production of the import good decreases. In Figure 15.7(a), agricultural production (the import good) decreases from

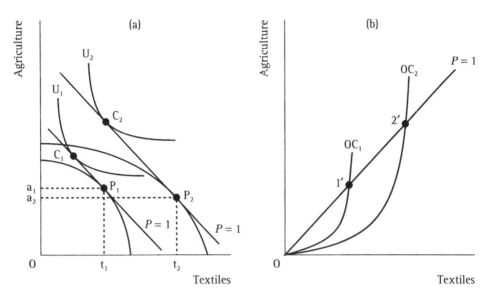

Figure 15.7 Growth and trade: the small-country case.

a_1 to a_2, while textile production (the export good) increases from t_1 to t_2. Thus the rate of increase for the export good is greater than that for the import good, and production is protrade. Consumption increases for both goods by approximately the same amount (Figure 15.7(b)). If the country's initial consumption of the export good is greater than its consumption of the import good, the consumption growth for the import good is proportionately more than that for the export good and consumption is protrade. The net effect is that this type of growth is protrade.

2 *The large-country scenario.* A similar example to that of the small country is examined in Figure 15.8. The underlying characteristics of the country and the type and degree of growth are identical to the previous example. The only exception is that this country is now large enough that changes in its production or consumption influence world prices. The PPFs are identical to those shown in the left panel of Figure 15.7.

In the large country, prices change and the offer curve of the rest of the world is included to show price determination. As in the small-country case, this country selects its initial level of production, P_1, that allows it to consume at point C_1 in the left panel and trade to the corresponding point $1'$ in the right panel. The initial world price of 1 ($P_T/P_A = P = 1$) is determined by the intersection of the country's offer curve OC_1 and the offer curve of the rest of the world OC_W.

In the small-country case, as the country's PPF shifts outward the country selects a production point P_2 that allows it to consume at point C_2 and trade to the corresponding point $2'$. However, this country's increase in its exports of the manufactured good depresses the world price from its initial level of 1 ($P_T/P_A = P = 1$) to $P = 2/3$, as determined by the intersection of the offer curves OC_2 and OC_W. Given this new

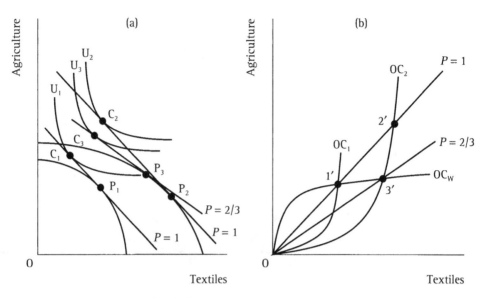

Figure 15.8 Growth and trade: the large-country case.

world price, the country will select the production point P_3 that allows it to consume at point C_3 and trade to the corresponding point $3'$.

The previous small-country example indicated that balanced-growth or growth biased toward export commodities will typically improve welfare. Although this scenario still shows a welfare improvement from the initial consumption point, the current example hints at the possibility that growth can actually result in welfare losses for a large country. Suppose that consumption point C_1 lay above the new PPF. If either the country's offer curve (OC_2) or the offer curve of the rest of the world (OC_W) were shaped in such a manner, it is possible that the resulting terms of trade could deteriorate to such an extent that the welfare associated with consumption point C_3 could be lower than the initial welfare level.

This situation in which the terms of trade erodes to the extent that it leads to a decline in the nation's welfare is referred to as *immiserizing growth*. This occurs despite the fact that the wealth effect typically increases the nation's welfare. While immiserizing growth may not be common, there are certain attributes associated with agriculture that make it quite possible, especially for developing countries. The likelihood of immiserizing growth increases when a nation is so large, in terms of the world market, that any increase in exports will deteriorate its terms of trade. Other factors that contribute to immiserizing growth include the rest of the world exhibiting very low income elasticities of demand for this product. Also, the nation relies so heavily on trade that a deterioration in its terms of trade results in lower national welfare. Given these conditions, it is not surprising that immiserizing growth is more common in developing countries.

■ 15.4 AGRICULTURE IN DEVELOPED AND DEVELOPING ECONOMIES

The role of the agricultural sector in an economy typically evolves as it moves through various stages of economic development. The agricultural sector in LDCs usually employs a large percentage of the workforce. In addition, a large percentage of the economy is involved in the production of food, and many households spend a significant amount of their available time producing their own food. With large percentages of their populations in rural areas, developing countries often experience difficulties providing access to quality education for all. Similarly, without development of the transportation and communication infrastructure, residents of farms and rural areas do not have the same access to the decision-making processes as their urban counterparts. As a result, agriculture is usually not protected, and in fact is often taxed, in LDCs.

The development of agriculture, especially during the twentieth century, has been shaped by labor-biased technical progress in the developed countries. The mechanization of agriculture and adoption of labor-biased herbicides and pesticides has lessened the need for labor in agriculture. As a result, industrialized nations such as the USA now employ less than 2% of their workforce in agriculture, yet still produce a significant portion of the world food supply. The reduced need for labor in agriculture allows for a greater degree of specialization within the economy. In a shift from a subsistence economy, individuals are able to achieve gains from specialization, resulting in productivity gains throughout the economy. The combination of a more productive workforce in agricultural production and increased productivity in other portions of the economy results in cheap food and higher incomes. As a result, developed countries typically spend a much smaller portion of their income on food.

As the population shifts toward urban and suburban areas, DCs experience less difficulty in providing access to quality education for their populations. At the same time, as the population tends to cluster around towns or cities, transportation and communication problems ease. But more than increasing access for the larger number of nonrural residents, increased national income results in a greater availability of resources to address those infrastructure issues of farms and rural areas. The improved transportation and communications infrastructure available to residents of rural areas allows for greater access to policy-makers and decision-centers. Agricultural producers in DCs are better able to organize than their LDC counterparts. This is one explanation of why the level of domestic support and protection given to agriculture in DCs is significantly higher than in LDCs.

■ 15.5 DEVELOPMENT ISSUES IN THE WTO

Agriculture plays an important role in the economic growth and development of the majority of LDCs. In recent years, this situation has become obvious with the demands of developing countries in the WTO. DCs have traditionally been the primary agents for determining the rules of multilateral trade agreements. As evidenced through the negotiating proposals from developing countries, there is little doubt that LDCs will attempt to use the WTO agricultural negotiations to obtain a greater degree of agricultural trade liberalization than achieved in the Uruguay Round of GATT. Specific objectives for trade negotiations as outlined by the LDCs include increased access to world markets for the agricultural exports of LDCs, more equitable conditions in the world market – particularly regarding export subsidies in the USA and EU – and demands for special treatment.

As the agricultural trade performance of LDCs has evolved, their interests in the current WTO negotiations have shifted. International trade in agriculture has increased in processed and other higher-value products. Concessions on agricultural raw materials, other bulk commodities, and tropical products now cover only a portion of developing country trade interests. Since LDCs are producing a large amount of temperate products, they seek market access and trade liberalization. Concessions by DCs within the WTO negotiations on temperate products produced by LDCs and on processed agricultural products will be a priority.

Import markets in DCs are not the only markets to be considered. The importance of LDCs and transition economies as outlets for agricultural goods will continue to increase in importance. The share of LDC agricultural exports going to other LDCs and transition economy markets has grown to approximately 40%. This indicates that these markets are almost as important to producers and processors in developing countries as the agricultural markets of the industrialized world.

LDCs are concerned with the impact of predatory and subsidized imports in their own markets. This can be seen in the proposals for special and differential treatment in terms of countermeasures permitted under the WTO. The Uruguay Round Agricultural Agreement created special agricultural safeguard mechanisms to provide protection for developing countries. One question to be dealt with in the WTO negotiations relates to whether these safeguard mechanisms should be utilized within the next agreement or whether distortions should be corrected at their source. While the answer to this may seem straightforward, there are developed countries whose positions are quite defensive toward maintaining the status quo with respect to their advantage with particular products in traditional markets.

A related issue for developing countries is the use of special and differential treatment in the form of extended preferential tariff treatment for agricultural products originating in developing countries. Once again, the question to be considered is whether the interests of developing country exporters are better served by expanding market access to all countries or through preferential market access arrangements. Similar to the earlier scenario, there are developing countries whose positions are quite defensive toward maintaining the status quo with respect to their advantage with certain products in certain traditional markets.

Specific targets that developing countries would like to achieve in the current negotiations include the further reduction of direct export subsidies and the creation of more effective disciplines on other forms of export subsidies. The overall welfare of both developing and developed agricultural exporting countries will increase when export prices are determined by the market rather than through government price

BOX 15.1 THE HISTORIC ROLE OF FOOD AID

Food aid is a mechanism that has often been used to alleviate hunger in low-income countries. It was first utilized by the USA in the early 1950s, as a means of disposing of excess cereal production. As can be seen in Table 15.1, the USA is by far the largest donor country or trading bloc in terms of cereal food aid, followed by the EU, Australia, Canada, and Japan.

Donor countries typically specify humanitarian relief and hunger as their primary criteria in determining food aid levels. However, political and economic factors often play a significant role in determining the amount and allocation of food aid. The effects of food aid are mixed. On one hand, food aid has the positive effect of increasing the supply of food when food shortages occur. However, food aid can also result in a production disincentive to domestic producers, since it depresses domestic prices in the recipient countries. Another negative aspect of food aid is that it is often based on the priorities of the donor countries, rather than on the needs of the recipient.

The USA provides food aid to a variety of countries through several mechanisms. While the primary objective of food aid is humanitarian, the provision of food aid can benefit domestic commodity producers in the donor country. Regardless of the reason why food aid is provided, gifts of food aid can have unintended negative consequences for the agricultural production sector of the country receiving aid. An increase in the supply of food, regardless of whether it is received as a gift or through credit programs, can depress prices in the local market. Profit-maximizing producers would react to these signals by decreasing production, exacerbating the shortage.

Table 15.1 Contributions of cereal food aid (1,000 tons)

Country	1995/96	1996/97	1997/98	1998/99	1999/00
Australia	181	170	296	267	264
Canada	436	373	384	332	349
EU	1,731	1,099	890	1,572	1,324
Japan	821	292	356	936	303
USA	3,037	2,273	2,787	6,390	6,693
Total	7,397	5,605	6,241	11,034	10,228

Source: Food and Agriculture Organization of the United Nations

support programs. Food importers are also realizing that subsidizing exports by their trading partners harms the development of their own agricultural sectors.

Another important issue in the agricultural negotiations from the perspective of developing countries concerns limitations on domestic support. The current agreement has allowed countries to use domestic support measures to address various non-trade concerns. These include issues such as poverty alleviation, rural development, food security, and environmental concerns. Despite this, many developing countries perceive the Uruguay Round agreement concerning domestic support as providing greater advantages to the industrialized world rather than to LDCs. It may be straightforward for countries to agree that clearer definitions are needed to determine the WTO legality of domestic support policies. While this may be the case, the more difficult challenge lies in convincing developed countries to limit their right to provide domestic agricultural support.

SUMMARY

1 There are several ways in which an economy can experience growth in its factors of production or endowments. An increase in the endowments of labor and capital causes the nation's production frontier to shift outward. The manner in which this shift occurs depends on the rate at which labor (L) and capital (K) grow. If L and K grow at the same rate, the nation's PPF will experience *balanced growth.*

2 The *Rybczynski theorem* states that at constant commodity prices, an increase in the endowment of one factor of production will increase the output of the good which uses that factor most intensively and will reduce the output of the other good.

3 Technical progress is classified as either endowment-neutral, labor-biased, or capital-biased. All types of technical progress, regardless of their classification, reduce the amount of inputs necessary to produce one unit of output.

4 The situation in which the terms of trade erodes to the extent that it leads to a decline in the nation's welfare is referred to as *immiserizing growth.* Immiserizing growth occurs despite the fact that the wealth effect typically serves to increase the nation's welfare on its own.

5 The role of the agricultural sector within an economy typically evolves as a country moves through the various stages of economic development. The agricultural sector in less developed countries usually employs a large percentage of the workforce. In addition, agriculture is usually not protected, and in fact is often disprotected, in developing countries.

6 The USA provides food aid to a variety of countries through several mechanisms. The major authorities for grant and concessional-credit food aid in the USA are Public Law 480, the Food for Progress Act of 1985, and Section 416(b) of the Agricultural Act of 1949.

KEY CONCEPTS

Balanced growth – A proportional increase in the factors of production capital.

Capital-biased technical progress – Increased productivity resulting to a greater degree from increased productivity of capital than for labor.

Endowment growth – An increase in the factors of production.

Endowment-neutral technical progress – Increased productivity resulting equally from the productivity of capital and the productivity of labor.

Food aid – A mechanism that has often been used to alleviate hunger in low-income countries.

Immiserizing growth – The situation in which the terms of trade erodes to the extent that it leads to a decline in the nation's welfare.

Labor-biased technical progress – Increased productivity resulting to a greater degree from increased productivity of labor than for capital.

Rybczynski theorem – The proposition that states that at constant commodity prices, an increase in the endowment of one factor of production increases the output of the good using that factor intensively and reduces the output of the other good.

Technical progress – Increased efficiency or productivity due to the development and adoption of more productive technologies.

QUESTIONS AND TASKS FOR REVIEW

1 Provide examples of capital-biased and labor-biased technical progress in agriculture. How does each of these affect the returns to labor and capital?

2 What implications does the Rybczynski theorem have for countries that wish to borrow capital to aid in their development?

3 What is immiserizing growth? What factors should countries take into account to avoid this phenomenon?

4 How can balanced growth result in the favoring of one commodity over another?

5 How can the humanitarian provision of food aid result in negative consequences to the recipient countries? What should donor countries do to avoid these results?

6 Describe the position of developing countries in the WTO agricultural negotiations.

SELECTED BIBLIOGRAPHY

Bhagwati, J. 1958: Immiserising growth, a geometrical note. *Review of Economic Studies*, XXV(3), 201–5.

Bhagwati, J., Jones, R., Mundel, R. and Vanek, J. (eds.) 1971: *Trade, Balance of Payments and Growth: Papers in International Economics in Honor of Charles P. Kindleberger.* Amsterdam: North-Holland.

Cochrane, W. 1979: *The Development of American Agriculture.* Minneapolis, MN: University of Minnesota.

Findlay, R. 1984: Growth and Development in Trade Models. In R. Jones and P. Kenen (eds.), *Handbook of International Economics*, vol. 1. Amsterdam: North-Holland.

Hayami, Y. and Ruttan, V. 1985: *Agricultural Development: An International Perspective.* Baltimore, MD: Johns Hopkins University Press.

Johnson, H. G. 1962: *Economic Development and International Trade in Money, Trade, and Economic Growth.* Cambridge, MA: Harvard University Press, 75–103.

Rybczynski, T. M. 1955: Factor endowment and relative commodity prices. *Economica*, XXII(84), 336–41.

Shapouri, S. and Rosen, S. 2001: *Food Security and Food Aid Distribution.* Agriculture Information Bulletin Number 765-4, USDA, Economic Research Service, Washington, DC, April.

APPENDIX 15.1 PROOF OF THE RYBCZYNSKI THEOREM

From the labor constraint equation (15.1), we obtain

$$A = \bar{L}/a_y - (a_x/a_y)T \tag{A15.1-1}$$

and from the capital constraints equation (15.2), we obtain

$$A = \bar{K}/b_y - (b_x/b_y)T \tag{A15.1-2}$$

Solving these two equations simultaneously,

$$T = (b_y\bar{L} - a_y\bar{K})/(a_xb_y - a_yb_x) \tag{A15.1-3}$$

$$A = (a_x\bar{K} - b_x\bar{L})/(a_xb_y - a_yb_x) \tag{A15.1-4}$$

If we assume that

$$a_x/a_y > b_x/b_y, \quad a_xb_y > a_yb_x \tag{A15.1-5}$$

then the denominator is positive. An increase in $\bar{L}$ increases A and decreases T, while an increase in $\bar{K}$ increases T and decreases A.

CHAPTER 16

Trade and the Environment

■ 16.0 INTRODUCTION

As countries go through the development process, the primary concern of individuals changes from meeting basic needs to achieving other goals. The position of countries with respect to the environment is no different. Most individuals in developed countries have little concern regarding the basic needs, such as food, shelter, and clothing. They can concern themselves with other issues, such as sustaining the quality of their environment. On the other hand, if individuals in less developed countries are worried about where their next meal will come from, they have little time to spend considering environmental issues. As a result, the developing world tends to discount future environmental quality in favor of more immediate needs. To avoid restrictive environmental rules and regulations, firms in developed countries (DCs) tend to move pollution-causing production facilities to less developed and developing countries (LDCs), who welcome the economic stimulation even though the production facilities create pollution.

The implications of these differences between countries for trade liberalization are profound. Negotiators attempt to identify an area of compromise that allows LDCs to achieve growth while being acceptable to DCs from an environmental perspective. The paradox is that a policy to achieve environmental protection could retard growth in LDCs. This does not allow the income of LDCs to grow to a level at which the environment increases in priority.

This conflict was apparent at the Seattle meetings of the WTO, which highlighted the linkages between international trade and other issues such as the environment and labor conditions. This chapter exposes students to the underlying issues regarding trade and the environment, comparing the economics of trade and the environment. The existence of environmental production externalities and their potential for sub-optimal outcomes are discussed. Various issues associated with the lack of harmonization of standards across countries are presented. The chapter concludes by discussing the environmental consequences of agricultural trade liberalization, as well as the role of environmental issues in achieving multilateral and regional trade agreements.

■ 16.1 INTERDEPENDENCE BETWEEN TRADE AND THE ENVIRONMENT

Countries trade with one another for a variety of reasons. Gains from trade occur when countries are able to focus their production activities in a way that takes advantage of their endowments, technologies, and other related factors. Trade also allows consumers to enjoy a greater choice of goods and services at lower prices. Although there may be groups that are harmed by trade, in general, trade is viewed as a means of benefitting the economy.

Despite the consensus that free trade is ideal, the movement toward free trade has raised concerns regarding potential conflict between trade and the environment. Growth and development benefit society by increasing the quantity of goods and services available for consumption. International trade is a vehicle for achieving this growth. However, as countries develop and as trade becomes freer, the concern arises as to whether the growth achieved is sustainable. In other words, is the level of current economic activity depleting or degrading the resource base in such a way as to negatively impact the well-being of future generations? Given that farmers in many countries have traditionally been known as stewards of the land, it is fitting that the interaction between agricultural trade and the environment be examined.

This interdependence is perhaps best stated in the founding charter of the WTO. In April 1994, the *Marrakesh Agreement Establishing the World Trade Organization* was signed. Its preamble stresses the importance of working toward sustainable development, when it states that WTO members recognize:

> that their relations in the field of trade and economic endeavour should be conducted with a view to raising standards of living, ensuring full employment and a large and steadily growing volume of real income and effective demand, and expanding the production of and trade in goods and services, while allowing for the optimal use of the world's resources in accordance with the objective of sustainable development, seeking both to protect and preserve the environment and to enhance the means for doing so in a manner consistent with their respective needs and concerns at different levels of economic development.

That the first paragraph of the preamble identifies sustainable development as a key component of the multilateral trading system signifies the value that WTO members place on the environment.

CONFLICT AND COMPATIBILITY BETWEEN TRADE AND THE ENVIRONMENT

As a rule, the WTO does not interfere with domestic measures to protect the environment. The WTO's primary concern has historically not been with the environment, but in the reduction of tariffs and the creation of freer trade. As a result, WTO rules may be incompatible with international environmental agreements or could interfere with the enforcement of environmental rules when trade is involved.

An example of the conflict between trade and the environment is the Dolphin–Tuna case between the United States and Mexico. Mexico challenged the validity of a US law that banned imports of tuna caught using techniques that kill dolphins. A 1991 GATT panel upheld the Mexican challenge. One of the reasons for this ruling was that GATT rules are concerned with the product being traded, not the techniques or processes by which the products are produced. The Dolphin–Tuna case was not an isolated incident. There were several areas in which environmental laws and regulations were in conflict with the GATT and later the WTO. As a result, environmental groups targeted the WTO as an environmentally hostile institution. This resolve was clearly evident at the Seattle meetings of the WTO, as environmental activists were able to slow the pace of the negotiations.

As the environment receives a greater degree of attention and importance in multilateral trade negotiations, it is likely that the conflicts between trade and the environment will increase rather than disappear. As trade volumes and environmental regulations increase, the incidence of environmentally related trade conflicts could become a larger portion of the cases considered in the WTO dispute settlement process.

One area in which conflict can arise between international trade and the environment involves the relationship between the environment and competitiveness. Environmental standards vary greatly among countries. While regulations tend to be less restrictive in LDCs, the divergence of environmental standards among DCs also tends to be quite high. Some countries argue that lower environmental standards provide an unfair advantage to industry in less regulated countries. Firms that must meet environmental standards in a regulated country face a higher cost of production than unregulated firms. From a cost perspective, firms in regulated countries lose competitiveness with respect to cost of production relative to unregulated firms.

Another area for conflict involves environmental policy jurisdiction. Environmental groups often call for action against countries that do not comply with their prescribed environmental behavior. An example of this would be a US environmental interest group boycotting Mexican tuna in an attempt to achieve a desirable outcome. While strategies of this type used by private groups may be quite successful, it is important to note that restrictive actions of this type by governments violate WTO rules.

A third area of potential conflict involves trade and the transnational spillover of pollution. Air and water pollution are two types of emissions that have the potential to cross national boundaries. When pollution crosses borders, one solution involves multilateral environmental agreements (MEA). However, there may be situations in which countries do not wish to participate in an MEA. Reasons for this may include uncertainty regarding the outcome due to a lack of scientific evidence, a perception that the proposed remedy would be ineffective, or stronger domestic regulations.

Perhaps the best known example of an MEA is the Kyoto Protocol. In 1992, the United Nations Framework Convention on Climate Change (UNFCCC) called for industrialized countries to voluntarily stabilize their emissions of greenhouse gases at 1990 levels by the year 2000. The Kyoto Protocol to the UNFCCC goes beyond the initial agreement by legally committing 38 major industrial countries to specific emission reductions. While the USA ratified the 1992 UNFCCC, the Bush Administration rejected the Kyoto Protocol in March, 2001, and indicated that it will seek new approaches based on voluntary measures and market mechanisms. Since then, other parties have proceeded to seek ratification of the agreement without the participation of the USA.

■ 16.2 ECONOMICS OF TRADE AND THE ENVIRONMENT

Agriculture is an industry with much potential for interaction with the environment. Much of this interplay takes the form of spillover benefits or consequences. Some governments support agricultural production activities as a way of enhancing the beauty of the countryside. At the same time, farming can harm the environment through air and water degradation. Given this, it is important to understand the economics of production externalities.

ECONOMICS OF PRODUCTION EXTERNALITIES

The primary objective of producers is typically to create and sell a product to maximize profits. However, the production process often results in byproducts that impact other members of society. These byproducts are known as externalities. An externality is the positive or negative effect of the actions of an economic agent on another without compensation. Positive externalities provide some type of benefit to the other entity. An example of a positive externality is when the honey bee pollinates the neighbor's crops while collecting nectar for the beekeeper. This is an externality if no fee is charged for the pollination. Other examples of positive externalities include enjoying the view of the neighbor's garden or admiring the rural landscape on a trip through the countryside.

An example of a negative externality is secondhand smoke. Many people enjoy smoking. However, people who do not enjoy secondhand smoke or who develop health problems because of it are not compensated by smokers. This is a negative externality. Air and water pollution are also examples of negative externalities.

By its nature, the agricultural industry has a number of potential negative externalities associated with the environment. Pesticides and livestock wastes can enter groundwater or surface water; livestock odors can drift to neighboring properties; soil erosion can damage crops and affect waterways; and agribusiness processors can generate water and air pollution.

In considering the role of externalities in agricultural trade and the environment, the focus tends to be on negative externalities. There are costs associated with negative externalities. At the same time a trade-off exists, in that regulations (at a cost) can

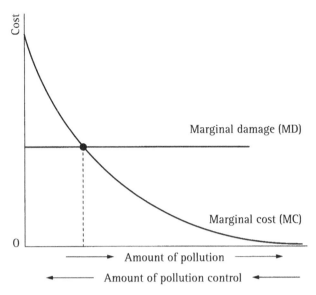

Figure 16.1 Marginal costs of control and marginal damages of pollution in the fertilizer industry.

reduce negative externalities. This trade-off between costs of damages and regulation are examined in Figure 16.1. Suppose that a fertilizer production plant produces a negative externality in the form of air pollution. The marginal damage curve (MD) indicates the additional damage done by each unit of pollution resulting from fertilizer production. For simplicity, we assume constant marginal damages. In other words, each additional unit of pollution increases the total environmental damage by the same amount. The marginal cost curve (MC) indicates the cost of removing one unit of pollution. The MC shows that when there is a large amount of pollution, the cost associated with decreasing the pollution level by one unit is rather low. The cost associated with removing one unit of pollution increases as the level of pollution declines, such that it is very expensive to eliminate the final unit of pollution.

The optimal level of pollution and regulation can be found at the point at which the marginal cost of regulation is equal to the marginal damages of pollution. This is the point at which MC is equal to MD. Note that any point to the right of this equilibrium results in excess pollution. In other words, the damages done by one additional unit of pollution are greater than the cost of reducing pollution. Any point to the left of the equilibrium results in excess regulation. Pollution decreases, and with it marginal damages from pollutants decrease. However, the cost of regulation now exceeds the gains from pollution reduction.

This example shows that neither the complete deregulation nor the total elimination of pollution would be the optimal solution. The best outcome is an intermediate point that results in a reduction, but not the complete elimination, of pollution.

ECONOMICS OF DIFFERING ENVIRONMENTAL STANDARDS

Once the optimal level of pollution has been determined, the government may choose to regulate the industry to achieve this outcome. This can be done through a variety of methods that can include penalties for polluting or rewards for compliance. Regardless of the technique chosen, successful regulation will cause the cost of production to increase, causing the supply curve of the industry to shift left.

Countries that choose to regulate their industries and protect their environment do so with the knowledge that there is a cost to the polluting industries associated with their actions. However, the issue becomes less clear when a country is involved in trade. Consider the trade impacts of differing environmental standards across countries in the two-country world presented in Figure 16.2, in which Mexico and the USA produce and consume fertilizer. In the initial situation, neither country imposes any regulatory standards on its fertilizer industry. An initial equilibrium world price of $3.50 occurs as a result of the intersection of Mexican export supply ES_M with US import demand ED_{US}. The USA imports 50 units of fertilizer from Mexico.

Now suppose that the USA chooses to regulate its fertilizer industry to limit pollutants that result from fertilizer production. To accomplish this, the US government imposes a tax on fertilizer production. This serves to increase the US cost of production, shifting the supply curve from S_{US} to S'_{US}. As a result, the US import demand curve shifts from ED_{US} to ED'_{US}. The new equilibrium world price of $4.50 occurs as a result of the intersection of Mexican export supply ES_M and the new US import demand ED'_{US}.

The USA increases its imports of fertilizer from 50 to 75 units. This may be of concern to US producers, consumers, and environmental groups, because this good is coming from a country without environmental regulations. At the same time, US production of fertilizer declines from 60 to 25 units, while US consumption of

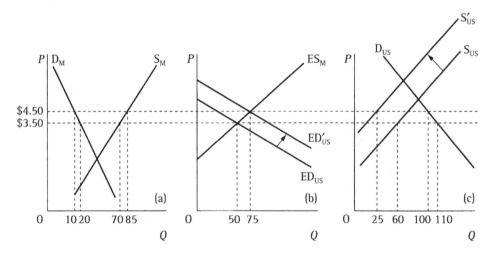

Figure 16.2 Trade effects of pollution control regulations in the fertilizer industry:
(a) Mexico; (b) the world; (c) the USA.

fertilizer declines from 110 to 100 units. In general, a tax on US fertilizer companies leads to higher costs and a decreased market share. The US fertilizer industry loses competitiveness with respect to the unregulated country, Mexico.

In this example, the USA has been successful in reducing pollution within its own borders. However, a large part of the reduction may have occurred by displacing pollution rather than reducing it. If Mexican fertilizer processors employ technologies that generate more pollution per unit of fertilizer relative to US processors, it would be the case that world pollution actually increases as a result of US regulation.

Another issue relates to the location and efficiency of production facilities within the nonregulated country. While stricter regulations in the USA may reduce the domestic production of pollutants, the migration of these production facilities could increase Mexico's pollution. In the case of two countries in close proximity, such as the USA and Mexico, Mexican processing facilities may be located along the US border. For example, maqilladoras, located in northern Mexico, take advantage of transportation efficiencies and low-cost Mexican labor as they produce goods that are often targeted toward the US market. If a significant number of these facilities are located close to the US border, increased production could generate additional pollution that may spill over into the USA. If these firms create more pollution per unit of output than their US counterparts, a scenario could result in which US environmental quality is even worse than it was prior to the imposition of stricter regulations.

■ 16.3 ENVIRONMENTAL CONSEQUENCES OF TRADE LIBERALIZATION

It may appear that the environmental consequences of trade liberalization are related more to the outcome of conflict as opposed to compatibility. However, trade liberalization has the potential to result in optimal environmental outcomes. Through comparative advantage, international trade improves the allocation of resources throughout the world. This enhances incomes in developing countries and increases their demand, not only for goods and services but for environmental amenities as well. Increased income levels provide incentives for countries, both developed and developing, to raise environmental standards.

As the interdependence between trade and the environment is examined, it becomes obvious that there are both positive and negative effects. These effects of trade liberalization on the environment can be divided into several areas: the technique effect; the scale effect; the composition effect; and the transportation effect.

As per capita income increases, society tends to call for regulations that require the use of more environmentally friendly production methods. Trade liberalization can have a technique effect as producers modify production techniques. New production technologies can be either environmentally friendly, neutral, or harmful. This change can occur for several reasons. For example, increases in income may result in stricter environmental restrictions. Changes in relative prices may also induce innovation and alter production technologies. It may also be the case that foreign direct investment makes cleaner technologies more widely available. An additional cause for this change is that the use of lower environmental standards designed to attract industry may increase pollution.

Evidence has shown a positive correlation between trade and economic growth. However, increased output and size of the firm or industry may add to pollution and exacerbate natural resource depletion. This is the scale effect. The opposite may hold if the economy shrinks. During the transition of formerly centrally planned economies to more open market-oriented economies, huge reductions in pollution emissions occurred as their manufacturing sectors collapsed.

Trade liberalization may also have a composition effect on the output of an economy. Resources previously utilized in protected inefficient industries will be shifted to more productive use. Phasing out of the Multi-Fibre Agreement under the Uruguay Round will cause textile production in developing countries to increase, while production in their heavy manufacturing sectors will decrease. Because textile production is a cleaner industry than heavy manufacturing, there may be a positive composition effect, but it could be dwarfed by a negative scale effect.

Increased trade results in an increase in the movement of goods across borders. It may also affect the way in which goods are transported. Different types of transportation are associated with different levels of pollution and the net transportation effect depends on the types of transportation used following trade liberalization.

BOX 16.1 HOW DOES TRADE AFFECT THE ENVIRONMENT?

Proponents of free trade argue that multilateral trade liberalization will shift food production away from developed countries and toward developing countries that utilize labor more intensively and chemicals less intensively. Because land use is not very responsive to price changes, this shift will not accelerate tropical deforestation. As a result of a more efficient allocation of world resources, trade liberalization will increase income. This increase in income will further reduce environmental degradation.

A recent OECD study indicates that trade liberalization would cause agricultural prices and production to decrease in countries that are using chemical inputs most intensively. Conversely, countries that are using low levels of pesticides and fertilizers would experience increased application rates. This study concludes that trade liberalization will result in small effects on agricultural land use.

Enhanced income resulting from trade liberalization may result in an increase in environmental awareness and support of regulation. It is clear that environmental concerns are recognized as a key motivation in multilateral trade negotiations. What remains to be seen is whether trade and the environment will coexist as compatible objectives or be plagued by the conflict of contradictory goals.

BOX 16.2 WTO RULES AND REGULATIONS PERTAINING TO THE ENVIRONMENT

The WTO is by definition a trade organization and it does not have an agreement dealing specifically with the environment. However, a number of GATT agreements have contained provisions that deal with the environment. In fact, the *Marrakesh Agreement Establishing the World Trade Organization* stresses the importance of working toward sustainable development and seeking to protect and preserve the environment.

In seeking to better incorporate these ideals, the WTO established its *Trade and Environment Committee*. This committee is charged with studying linkages between trade and the environment, and making recommendations as to how trade agreements should be structured on the basis of these relationships. The work of the committee is governed by two basic principles. First, the WTO is a trade organization and is not suited to intervene in environmental policy issues. Its task is to study environmental policies only when these policies have a significant impact on international trade. Second, when conflicts arise between the WTO and environmental policies, solutions must be consistent with the basic principles of the WTO that an open, equitable, and nondiscriminatory trading system creates the best infrastructure for protecting and conserving environmental resources and promoting sustainable development.

Although the WTO is a trade organization and not an environmental organization, there are a number of WTO agreements that contain environmental provisions. Among these are the following:

- *General Agreement on Tariffs and Trade Article 20*: policies for protecting human, animal, or plant life or health are exempt from normal GATT disciplines under certain conditions.
- *Technical Barriers to Trade* (product and industrial standards), and *Sanitary and Phytosanitary Measures* (animal and plant health and hygiene): there is explicit recognition of environmental objectives.
- *Agriculture*: environmental programs are exempt from cuts in subsidies.
- *Subsidies and Countervail*: subsidies of up to 20% of firms' costs for adapting to new environmental laws are allowed.
- *Intellectual Property*: governments can refuse to issue patents that threaten human, animal, or plant life or health, or risk serious damage to the environment.
- *General Agreement on Trade in Services, Article 14*: policies for protecting human, animal, or plant life or health are exempt from normal GATS disciplines under certain conditions.

BOX 16.3 THE KYOTO PROTOCOL AND AGRICULTURE

There are currently several hundred multilateral environmental agreements that deal with environmental issues. Among these, the United Nations Framework Convention on Climate Change (UNFCCC) was enacted to address issues related to global climate change. An objective of the convention is to stabilize concentrations of greenhouse gasses (including carbon dioxide, methane, nitrous oxide, hydrofluorocarbons, perfluorocarbons, and sulfur hexafluoride) at safe levels. In 1992, the UNFCCC called for a voluntarily limitation of greenhouse gas emissions by the year 2000.

The Kyoto Protocol goes beyond the initial UNFCCC agreement in calling for legally binding commitments on 38 major industrial countries to specific emission reductions. Although the USA agreed to the initial UNFCCC, it rejected the Kyoto Protocol in March, 2001. The Bush Administration indicated that it will seek new approaches to accomplish greenhouse gas emission reductions that are voluntary and based on market mechanisms. Given the US stance on this issue, other countries have continued to complete the agreement without the formal participation of the USA.

The USA has indicated that it is in favor of a voluntary framework that includes emissions trading and developing country participation. One of the mechanisms called for in the Kyoto Protocol is the carbon charge. A carbon charge is a tax based on carbon emissions. Given the dependence of US agriculture on energy, many feel that a carbon charge will put US producers at a competitive disadvantage relative to developing country producers. This is especially relevant since the Kyoto Protocol places no restrictions on the greenhouse gas emissions of developing countries.

Because agricultural producers in developed countries face strong competition from their developing country counterparts, the standards called for in the Kyoto Protocol will have significant implications for agricultural trade. The impact of this type of agreement on agricultural trade can be seen by looking at developing countries who are major agricultural exporters. Argentina is a major exporter of wheat, coarse grains, corn, soybeans, and cotton. Brazil is a dominant soybean exporter. Thailand and Vietnam both supply a large portion of the world rice market, and India has a significant impact on the world cotton market. Imposing a carbon charge only on developed country production will increase the cost of developed country production relative to production in developing countries, potentially shifting the competitive positions of developed and developing countries. On the other hand, the competitiveness impacts on farmers in developed countries could vary relative to those in developing countries if the Protocol contains provisions to compensate farmers for environmentally friendly production practices.

BOX 16.4 ENVIRONMENTAL DISPUTES AND THE WTO: THE SHRIMP—TURTLE CASE

As trade agreements and environmental policy expand and overlap, it becomes difficult to determine which agreement has precedence or who should settle disputes. There have been several recent cases of trade disputes related to environmental issues. The following case provides an example of how trade and environmental agreements can conflict and how such disputes are resolved.

The US Endangered Species Act of 1973 lists five species of sea turtles that live within US waters as endangered or threatened. It prohibits hunting of these animals within the USA, its territories, or on the high seas. This Act requires US shrimp trawlers to use turtle-excluder devices when fishing in areas where sea turtles are likely to be found. To further its objective of protecting sea turtles, the USA enacted legislation restricting imports of shrimp harvested using techniques deemed detrimental to sea turtles.

In 1997, India, Malaysia, Pakistan, and Thailand filed a WTO complaint against the USA concerning the ban on imports of shrimp and shrimp products. In its ruling, the WTO stated that countries have the right to take trade action to protect the environment, including endangered species. It also stated that measures to protect sea turtles are acceptable under GATT rules, provided that these measures are nondiscriminatory.

The USA lost this case because it provided some countries, mainly in the Caribbean, with assistance and additional time to adopt approved harvesting techniques. India, Malaysia, Pakistan, and Thailand were not given the same advantages as certain countries in the Western Hemisphere. The Appellate Body stated that "this measure has been applied by the United States in a manner which constitutes arbitrary and unjustifiable discrimination between Members of the WTO, . . . , WTO Members are free to adopt their own policies aimed at protecting the environment as long as, in so doing, they fulfill their obligations and respect the rights of other members under the WTO Agreement."

SUMMARY

1 There are three general areas of conflict between international trade and the environment: (1) conflicts between the environment and competitiveness; (2) conflicts over environmental policy jurisdiction; and (3) trade and the transborder spillover of pollution.

2 An externality is the positive or negative effect of the actions of an entity on another, without compensation. Positive externalities provide some type of benefit to the other entity. Negative externalities cause damage to the other entity.

3 The optimal level of regulation can be found at the point at which the marginal cost of regulation is equal to the marginal damages done by pollution, $MC = MD$. Neither the

complete deregulation nor the total elimination of pollution is likely to be the optimal solution. The best outcome will likely be an intermediate point that results in a reduction, but not the complete elimination, of pollution.

4   Successful regulation will likely result in a leftward shift in the supply curve; the cost of production in a less-polluting manner will increase. The protecting country increases its imports of goods from the nonprotecting country. This may be of concern because this good is coming from a country without environmental regulations.

5   As the interdependence between trade and the environment is examined, there are both positive and negative effects. These effects of trade liberalization on the environment can be categorized into several different effects: the technique effect; the scale effect; the composition effect; and the transportation effect.

KEY CONCEPTS

Competitiveness – The ability to deliver a particular quality of goods to consumers at prices equal to or lower than competitors.

Composition effect – Causing resources previously utilized in various industries to be shifted to other industries with differing levels of efficiency.

Environmental policy jurisdiction – The attempt to alter various laws or operating procedures through action against those that do not comply with their prescribed environmental behavior.

Negative externalities – The adverse effects of the actions of an economic agent on another without compensation.

Positive externalities – The beneficial effect of the actions of an economic agent on another without compensation.

Scale effect – Causing increased output and size of the firm or industry may add to pollution and exacerbate natural resource depletion.

Technique effect – Causing producers to modify production practices.

Transborder spillover – The situation in which various types of emissions, such as air or water, cross national boundaries.

Transportation effect – Altering the type and amount of transportation which, in turn, impacts the pollution levels.

QUESTIONS AND TASKS FOR REVIEW

1   What is the role of the WTO concerning environmental issues? What is the environmental significance of the *Marrakesh Agreement Establishing the World Trade Organization*?

2   To what extent is there a conflict between trade and the environment? To what extent is there compatibility between trade and the environment?

3   Define externalities. What is a positive externality? What is a negative externality?

4   Show how regulations designed to correct for an externality can distort trade.

5   Discuss how trade liberalization can degrade the environment. Discuss how trade liberalization can improve the environment.

6   Identify the four types of effects that trade liberalization can have on the environment. Discuss each one. Indicate which are positive, which are negative, and which can be both positive and negative.

SELECTED BIBLIOGRAPHY

Anderson, K. and Blackhurst, R. (eds.) 1992: *The Greening of World Trade Issues*. Ann Arbor, MI: The University of Michigan Press.

Bredahl, M., Ballenger, N., Dunmore, J. and Roe, T. (eds.) 1996: *Agriculture, Trade, and the Environment: Discovering and Measuring the Critical Linkages*. Boulder, CO: Westview Press.

OECD (Organization for Economic Cooperation and Development) 2000: *Domestic and International Environmental Impacts of Agricultural Trade Liberalization*. Document # COM/AGR/CA/ENV/ EPOC(99)72/REV2. Paris: OECD.

Rao, P. 2000: *The World Trade Organization and the Environment*. New York: St. Martin's Press.

Rutan, V. (ed.) 1994: *Agriculture, Environment, and Health: Sustainable Development in the 21st Century*. Minneapolis, MN: University of Minnesota Press.

WTO (World Trade Organization) n.d.: Understanding the WTO environment: a new high profile. Available at http://www.wto.org/english/thewto_e/whatis_e/tif_e/bey2_e.htm

Index

Printed and bound by CPI Group (UK) Ltd, Croydon, CR0 4YY

16/04/2025

14658504-0005